Reactivity and Structure
Concepts in Organic Chemistry

Volume 23

Editors:

Klaus Hafner Jean-Marie Lehn
Charles W. Rees P. von Ragué Schleyer
Barry M. Trost Rudolf Zahradník

Karl-Dietrich Gundermann · Frank McCapra

Chemiluminescence in Organic Chemistry

With 30 Figures

Springer-Verlag
Berlin Heidelberg New York
London Paris Tokyo

Professor Dr. Karl-Dietrich Gundermann

Institut für Organische Chemie
Technische Universität Clausthal, D-3392 Clausthal-Zellerfeld

Professor Frank McCapra

The University of Sussex
The School of Chemistry and Molecular Sciences
Falmer, Brighton BN1 9QJ, United Kingdom

List of Editors

Professor Dr. Klaus Hafner
Institut für Organische Chemie der TH Darmstadt
Petersenstr. 15, D-6100 Darmstadt

Professor Dr. Jean-Marie Lehn
Institut de Chimie, Université de Strasbourg
1, rue Blaise Pascal, B.P. 296/R8, F-67008 Strasbourg-Cedex

Professor Dr. Charles W. Rees, F. R. S. Hofmann
Professor of Organic Chemistry, Department of Chemistry
Imperial College of Science and Technology
South Kensington, London SW7 2AY, England

Professor Dr. Paul v. Ragué Schleyer
Lehrstuhl für Organische Chemie der Universität Erlangen-Nürnberg
Henkestr. 42, D-8520 Erlangen

Professor Barry M. Trost
Department of Chemistry, The University of Wisconsin
1101 University Avenue, Madison, Wisconsin 53 706, U.S.A.

Professor Dr. Rudolf Zahradník
Tschechoslowakische Akademie der Wissenschaften
J.-Heyrovský-Institut für Physikal. Chemie und Elektrochemie
Máchova 7, 121 38 Praha 2, C.S.S.R.

ISBN 3-540-17155-X Springer-Verlag Berlin Heidelberg New York Tokyo
ISBN 0-387-17155-X Springer-Verlag New York Heidelberg Berlin Tokyo

Library of Congress Cataloging in Publication Data
Gundermann, Karl-Dietrich, 1922– Chemiluminescence in organic chemistry. (Reactivity and structure : concepts in organic chemistry ; v. 23) Includes indexes. 1. Chemiluminescence. I. McCapra, Frank, 1934– . II. Title. III. Series: Reactivity and structure ; v. 23. QD505.8.G86   1987   541.3   86–29722
ISBN 0-387-17155-X (U.S.)

© Springer-Verlag Berlin Heidelberg 1987
Printed in Germany

The use of registered names, trademarks, etc. in this publication does not imply, even in the absence of a specific statement, that such names are exempt from the relevant protective laws and regulations and therefore free for general use.
Typesetting: Friedrich Pustet, Regensburg. Printing: Mercedes-Druck, Berlin. Bookbinding: Lüderitz & Bauer, Berlin. 2152/3020-543210

Preface

The appearance of the first review in 1965 [1] and the first monograph in 1968 [2] on chemiluminescence demonstrated the extent of the phenomenon of light emission from the reaction of organic compounds in solution. Since then the number of chemiluminescent compounds has greatly increased, although the advances in theory and, more recently, applications are probably more significant.

The present work is written by two authors who, together with E. H. White, helped to bring the study of chemiluminescence into the modern era. However many investigators are making contributions to the subject, even if the number of enthusiasts still remains small.

It is not our intention to write an exhaustive account of chemiluminescence, still less of bioluminescence, and we have concentrated on making the landmarks in the area familiar to a readership outside the circle of specialists. The emphasis is on the range of organic compounds showing light emission with very little description of the relatively few inorganic or the more numerous biological examples which have been discovered.

We hope that some of the excitement of the striking demonstrations of chemiluminescence which can be made appears in the text, albeit in the form of intellectual satisfaction and interest.

We thank Prof. Dr. J. Stauff, Frankfurt for his generous advice and his critical comments. The chapter dealing with Peroxy-oxalate chemiluminescence has been commented up on critically by Dr. M. M. Rauhut, Stamford, Connecticut which we gratefully acknowledge.

Dr. Karl Wulff, Biochemical Research Center of Boehringer-Mannheim at Tutzing, was kind enough to help us find the final form of Chapter XIII (Analytical Applications . . .). We value this assistance very much.

The very valuable assistance of Dr. H. Fiege, Leverkusen is also highly appreciated. Dr. M. Steinfatt, Clausthal-Zellerfeld, has been excellently helpful, especially in the proofreadings. We are very much obliged to Dr. F. L. Boschke and his successor Dr. R. Stumpe, of Springer-Verlag, for their encouragement and patience without which this book would never have been written.

We have been helped in our work over the years by a large number of students and associates, whose names appear in the references. Many contributions have been made by them, and we take this opportunity of thanking them all most warmly. The appearance of this volume is also the direct result of the work of Mrs. S. Radecke, Mrs. I. Mielordt, Mrs. M. Th. Eckhardt and Mrs. C. Cundall who have been indefatigable and very able in transforming the manuscript into its final form. Mrs. R. Müller-Link, Mr. K. Mielordt and Mr. M.-R. Fischer skilfully produced the figures and formulas.

Clausthal-Zellerfeld, September 1986   K.-D. Gundermann
Brighton, September 1986   F. McCapra

Table of Contents

I. Introduction

Modern scientific research into the non-enzymatically catalyzed chemilumines-cence of organic compounds began with Radziszewski's investigations, in 1871, of the oxidation of lophine (2,4,5-triphenyl imidazole (1) by aqueous-alkaline hydrogen peroxide) [3].

Yellow light is emitted.

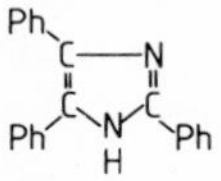

I (1) 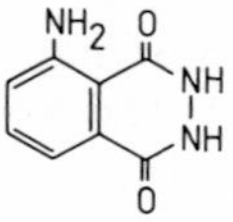   I (2) 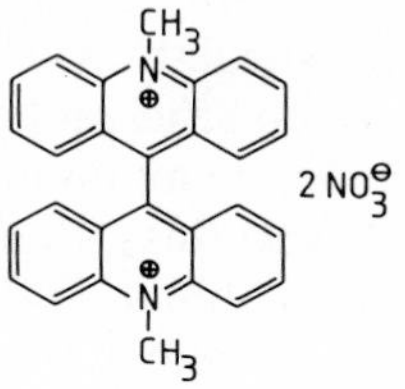   I (3)

This observation, accompanied as it was by knowledge of the structure of the compound involved, can be said to be the first report of the chemiluminescence of *organic* compounds in the modern sense. Radziszewski also described [4] the chemiluminescence which resulted from the oxidation of a wide variety of organic compounds and natural materials, indicating the generality of the reaction. The complexity of these prevented further progress, and the readily repeated observa-tions often still elude proper description in structural terms. Over fifty years passed before the best known organic chemiluminescent compound was disco-vered: Albrecht [5] in 1928 described chemiluminescence from luminol (2) under various conditions and made suggestions about the mechanism which still have to be taken seriously today.

Another isolated discovery in this sporadic progress was lucigenin [6] (3).

Dubois [7] in 1885 had shown that bioluminescence also involved the oxida-tion of a relatively small organic molecule, although this initiative also lay fallow for over fifty years.

The study of both bioluminescence and chemiluminescence which led to our present understanding was inaugurated by Harvey [8] in 1953 with the publication of "Bioluminescence". Also about this time White [9] made the first thorough and convincing study of the mechanism of luminol chemiluminescence. A review of the chemistry of the hydrazides, with special reference to their chemiluminesc-ence [10] also added to the foundations.

There are three main strands in this work. Increased interest in biolumines-cence, both from the biological and chemical standpoints, discovery of new efficient chemiluminescent compounds, and the creation of unifying theories

1

have all resulted in a satisfying whole. It is at present too early to assess the place of the applications of chemiluminescence, but the indications are very promising.

There is no single theory of organic chemiluminescence. Investigation of each of the well-defined systems must proceed in two stages. In the first the products of the reaction and the necessary conditions are characterised, and a reaction pathway described with as much precision as possible. The second phase, which includes the excitation step, is based on these findings. The physical process of excited state formation, by virtue of its transient nature (perhaps occurring within 10^{-15} s) is often merely inferred. Much depends on detailed knowledge of the chemical intermediate whose decomposition gives rise to excitation. On the other hand radiation from the excited state can be simply explained as fluorescence or, less often, phosphorescence.

Thus a recurring theme in the book is the search for the key intermediate, and major advances have resulted when it has been correctly indicated or identified.

For example, the suggestion [11] that dioxetans (Chap. V) would, on theoretical grounds, give rise to efficient population of excited states helped rationalize a large part of organic chemiluminescence and bioluminescence. The proposal stimulated the first synthesis [12] of what turned out to be a series of compounds of remarkably varied stability. Differences in substitution result in compounds so unstable as to defy isolation, whereas others are stable until their very high melting points are reached. The ease with which they can be modified and studied as the intermediates which give rise to excitation is a contribution which should not be underestimated. A similar advance resulted in a still more general mechanism of oxidation chemiluminescence – that of electron transfer [13] (Chap. IX). The centre – piece here is the active oxalate ester system, the most efficient of all synthetic chemiluminescence compounds [14].

Paradoxically, luminol and the cyclic hydrazides, which earlier looked set for an equally complete mechanistic understanding, have not yet reached that position. The main reason for this fact is the far more complicated chemistry of the luminol reaction. The modern work of E. H. White and Gundermann has provided a very solid basis for an acceptable description of the excitation step. However, convincing proof of the key intermediate appears to be lacking. Some recent work has nevertheless brought this most intriguing system within reach.

Other reactions which do not yield such visibly efficient chemiluminescence played their part in simultaneous advances. Fundamental work on singlet oxygen also appeared during the period, and has assisted in elucidating some of the weaker examples of luminescence [15, 15a]. Its involvement in less well defined biologically derived systems is however often still controversial.

The decomposition of peroxides generally gives rise to light emission, but those without special features are not efficient. The low level of light emission makes certain identification of mechanism difficult. This is particularly the case for very low level light from biological sources such as humic acid, brown coal oxidation and phagocytosis.

The rapid development of the chemiluminescence field is reflected in the number of conferences devoted entirely to the discussion of both chemi- and bioluminescence. The lively interest generated can be readily appreciated from the symposium volumes which have arisen [16, 17, 18, 19, 20].

The increasing significance of chemi- and bioluminescence in their clinical and biological applications has also been stressed in several international conferences [18, 19, 20].

Reviews which survey the field in a general way are appearing regularly [20, 21–30], and reference to more specific reviews is made in later sections.

A. Physicochemical Background

II. General Concepts in Chemiluminescence

Detailed descriptions of the various reaction mechanisms which have been identified will be given later at the appropriate points in the discussion. There are, however, several features which are common to all chemiluminescent reactions. These serve as a useful general introduction to most aspects of the phenomenon.

Chemiluminescence can be defined as the emission of light as a result of the generation of electronically excited states formed as a result of a chemical reaction. This occurs at temperatures below those necessary for incandescence, and the definition can be extended to include the reactions occurring in "cool flames" such as those of CS_2 and CO on reaction with O_2.

The range of wavelengths of light emitted is surprisingly large – from the near ultraviolet to the infra-red. Only a few reactions achieve sufficient energy to produce light with a maximum at wavelengths below 400 nm, and infra-red emission, while not being energy limited is even more rare in solution. A notable exception is that of certain fluorescers excited by active oxalates [31]. Several inorganic systems and gas phase atom recombination reactions are not specifically dealt with, although they lend themselves more readily to detailed physical examination. Some of the concepts thus generated are however used as a background to the study of the organic reactions. The discussion in this volume is thus essentially of the generation of visible light by organic chemical reactions in solution.

II.1 Chemiluminescence Efficiency

The efficiency of a luminescent reaction is defined as the number of photons emitted per reacting molecule, in einsteins. This quantum yield is usually written as

$$\emptyset_{CL} = \emptyset_R \times \emptyset_{ES} \times \emptyset_F$$

where $\emptyset_R$ is the yield of product, $\emptyset_{ES}$ the number of molecules entering the excited state and $\emptyset_F$ is the fluorescence quantum yield.

High yields of excitation without strong visible emission are possible if $\emptyset_F$ is very low, as is the case for simple dioxetans. Phosphorescence is difficult to observe under the conditions of most of the reactions, but in principle $\emptyset_{Phos}$ can replace $\emptyset_F$.

When the first formed excited state is quenched by a fluorescent molecule present in solution, with radiation from the acceptor, *sensitised chemilumines-cence* results. This form of emission is most easily seen when the transfer of energy takes place between singlet states (i. e. by the resonance or Förster mechanism

7

see Chap. II.2). However special features often facilitate the much less likely triplet to singlet transfer, and examples of both are discussed later.

The efficiency of this energy transfer (ET) must be taken into account:

$$\emptyset_{CL} = \emptyset_R \times \emptyset_{ES} \times \emptyset_{ET} \times \emptyset_{F'}$$

where $\emptyset_{F'}$ is the fluorescence efficiency of the acceptor, and $\emptyset_{ET}$ is the efficiency of energy transfer.

The range of efficiencies in chemiluminescence is great. The fact that a reaction is readily visible is no guarantee that efficiency is high.

For example the light emitted during the oxidation of a solution of *p*-chlorophenyl magnesium bromide is easily demonstrated, but $\emptyset_{CL}$ is only about 10^{-8}. On the other hand the decomposition of dioxetans can give excited states in greater than 50% yield, yet not appear as a bright reaction. In the first case it is probable that only a small fraction of the reactants undergo chemiluminescent reaction ($\emptyset_R$ is small). In many cases of weak chemiluminescence the fraction may be so small as to make identification of a chemical path impossible. Thus the reactions responsible for light emission during phagocytosis (Chap. XIII. A. 4.) can only be guessed at for the present.

Addition of fluorescent acceptors can enhance light emission in the case of both dioxetans and phagocytosis (i. e. by sensitisation), obtaining evidence for the yields of poorly radiating primary excited states.

II.1.1 Energetics of the Reactions

Since the emission is quantised, the reaction yielding the excitation must deliver, *in one step,* energy equivalent to the shortest wavelength observed. Visible light corresponds to energies of 168–294 kJmol^{-1} (40–70 kcal mol^{-1}). The thermochemistry of the reactions is rarely known, although measurements have been made in a few cases. Calculations using published bond energies and tables [32] give a useful estimate for those cases in which bond cleavage is involved. More accurate values can be obtained for electron transfer reactions using redox potentials obtained electrochemically. Thus it is possible to identify with some precision the energy released during electron transfer chemiluminescence, including electroluminescent reactions [33].

The correspondence between the energy of the 0–0 band of the emission spectrum and the energy released need not be exact since small amounts of thermal energy can contribute to reaching the level of the excited state: kT at room temperature is about 2,5 kJmol^{-1}. There is evidence [34] that this can occur in some reactions. It is therefore possible to exclude a putative chemiluminescent reaction on the grounds of energy deficiency. Alternatively secondary reactions, such as triplet-triplet annihilation can be indicated where light is emitted, but where there is insufficient energy to populate the singlet excited state directly [35, 36]. If the excitation step is preceded by changes in bonding, it is much more difficult to be sufficiently precise in these calculations.

II.1.2 Formation of the Excited State

Our understanding of the mechanisms of chemiluminescence has advanced to the point that some prediction of light emission from a chemical reaction is now possible. This can be done by reference to the various structural types and general mechanisms which have been established. However much remains to be done in identifying the fundamental physical processes which lead to electronic excitation. It is thus not possible to give a single answer to the question "what are the factors controlling excited state formation?", although electron transfer phenomena come fairly close. In spite of the difficulty, it is worthwhile summarising the various features which play a part.

II.1.2.1 The Reaction Co-ordinate

Chemiluminescence must be the result of a non-adiabatic reaction in that the ground state potential energy surface describing the molecule undergoing reaction must intersect with a different, excited state surface.

Potential energy surfaces for the reactions can be obtained with varying degrees of precision. Some are merely illustrative, [37, 38, 39] while others lend themselves to calculation [40, 41]. Among the factors which may influence the population of the excited state are the following.

II.1.2.2 Molecular Geometry

Although the Franck-Condon principle which governs the geometry of the first formed excited state in an electronic transition does not strictly apply to chemical reactions, a similar constraint is expected. The geometry of the reactant will not change during the instant when the excited state is formed. If the nuclear configuration of the transition state between reactant and (excited) product resembles the latter, then chemiluminescence is made more probable. Alternatively, it can be said that the geometry at the crossing point between ground and excited state surfaces (see later) favours entry into the excited state. A related consideration is the disposition of the free energy contained in the reactant in the short time scale of either a surface crossing or an electron transfer. Thus it has been suggested [42] that highly exothermic reactions producing small molecules are more likely to result in excited state formation since the energy would be difficult to accommodate in the limited degrees of vibrational freedom of such products.

II.1.2.3 Spin Restrictions

One of the most certain ways in which an excited state may be expected is by the operation of the conservation of spin [43]. If for example these rules require that triplet products are most probable, then excitation is likely since almost all organic molecules have singlet ground states. Although this concept commonly applies only to oxygen and nitroso-compounds (there are no chemiluminescent reactions known of the latter sort) among stable molecules, it seems to be important for the reaction of fragments and unstable intermediates (e. g. from dioxetans and hyponitrite esters).

The chemical production of singlet O_2 from a variety of singlet precursors does in fact produce excited states [44] since ground state oxygen is a triplet. Other (gas phase) reactions of small molecules, producing NO_2 and SO_2, can be shown to reflect this idea also.

Ph—O—O—Ph ... Ph—O—O—Ph

$\underline{I}(4)$ $\underline{I}(5)$

A concerted organic reaction in which O_2 is produced will also be controlled by the unique multiplicity of the oxygen ground state. For example, the two peroxides (4) and (5), discussed in detail later, must produce (if the decomposition is concerted) either excited (singlet) O_2 or excited (triplet) carbonyl products to conform to the spin rules.

II. 1.2.4 Symmetry Restrictions

More general orbital symmetry requirements might be expected to affect the probability of entering an excited state. A good example [45] is that of the formation of excited SO_2 and NO_2 discussed below. The relative ease with which the kinetic parameters may be obtained for such simple reactions allows quantitative confirmation of the importance of this effect. The other reaction for which such factors have been discussed is that of the dioxetans. It is more difficult to produce convincing evidence of the operation of symmetry rules in this case, although many calculations have been performed in pursuit of the idea [37–41]. A more complete discussion appears in the section dealing with the dioxetans (Chap. V).

II. 1.2.5 Potential Energy Surfaces

All of the above considerations can be illustrated in the course of a discussion of the potential energy surfaces of ground and excited state, and their interaction.

A particularly clear example is that of the reactions of SO and NO with O_3 studied by Thrush [45].

$$SO + O_3 \rightarrow SO^*_2 + O_2$$
$$NO + O_3 \rightarrow NO^*_2 + O_2$$

Both of these reactions are chemiluminescent and the potential energy surfaces which result in either ground state or excited state formation are shown.

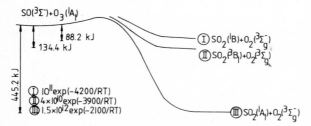

Fig. 1. Potential Energy Surfaces of the $SO + O_3$ – Reaction (after B. A. Thrush [45]).

The formation of excited NO_2 is particularly interesting in that the geometry of the approach of the reactants can be correlated with the product symmetry, the excited state being linear and the ground state bent. The differing probability of formation of each of the states can be demonstrated very nicely by the correlation between the values of the pre-exponential (log A) factors and the continuity of the PE surfaces as determined by the state symmetry.

The decomposition of dioxetans is a rich field in which various theories of excited state formation flourish. Indeed it is likely that various mechanisms such as electron transfer, radical (and hence triplet) re-combination and symmetry controlled reactions are all involved. Detailed discussion is best deferred but the principle involved is that the ground state of the dioxetan has the same symmetry as that of a doubly excited state of the carbonyl products. Thus a crossing of this potential energy surface with a lower energy excited state is likely. It might have been expected that crossing into a singlet excited state was most probable, and it is therefore surprising to find that for simple dioxetans, triplet products are almost exclusively formed [46]. This preference is in agreement with Richardson's view that cleavage of the O–O bond occurs first, and that it is the resulting diradical that leads to excited products [47]. The relaxation of this diradical from the initial singlet to the lower energy triplet is thus an example of a spin restriction favouring subsequent excited (triplet) state formation. Several calculations have examined both routes to excited carbonyl compounds [37–41].

Another example, discussed with the dioxetans later, which involves cage reaction of triplet radicals is that of the decomposition of hyponitrite esters [48]. Hydrogen abstraction from one radical $R_2CH–O\cdot$ by another leads to triplet carbonyl products in high yield.

All of the reactions discussed above produce excited carbonyl products which are expected [49] to have a partially tetrahedral configuration about the trigonal C atom rather than the planar geometry of the ground state carbonyl. Since these products are derived from tetrahedral sp^3 precursors, geometrical influences may be at work. The assessment of the relative worth of each of these factors in determining whether or not excited states are formed is not possible at present.

II.2 Electron Transfer Reactions

The strongly exothermic transfer of electrons between fluorescent organic molecules represents one of the most general mechanisms in chemiluminescence [50, 51]. It can be found in electroluminescence, radical ion annihilation and peroxide decomposition. The basic concept was introduced by Hercules [39] following the general theory of Marcus [52]. The reaction co-ordinate can be roughly indicated by the potential energy curves shown.

In the case of electron transfer, where geometry changes are relatively small, the overall activation energy is large where ΔH_r is also large. Thus the dissipation of the energy of a highly exothermic reaction occurs with a high activation energy if the ground state is formed directly. Formation of a more nearly iso-energetic excited state has a much smaller activation energy.

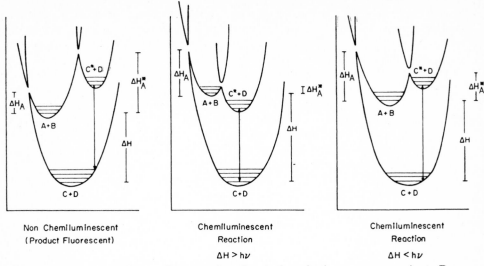

Fig. 2. Reaction coordinate diagrams for chemical excitation processes: A + B = reactants; C + D = products formed in ground states; C* + D = products formed with C in an excited state and D in the ground state; ΔH = energy available from the reaction according to the usual thermodynamic criteria; $\Delta H \neq$ = activation energy for formation of products in the ground state; $\Delta H \neq^*$ = activation energy for formation of one product in an excited state; $h\nu$ = energy necessary for the excitation C → C* (after D. M. Hercules [39]).

Originally invoked for electroluminescent reactions, this idea has now been developed for the reaction of peroxides with fluorescent compounds of low ionisation potential [51]. Many of these reactions are discussed in Chap. XI, but Fig. 2 can be most succinctly exemplified by the radical ion annihilation shown, where Ar is a fluorescent aromatic hydrocarbon such as diphenylanthracene, LUMO is the lowest, normally unoccupied molecular orbital and HOMO is the highest occupied molecular orbital.

$$Ar^{(-)} \quad + \quad Ar^{(+)} \quad \rightarrow \quad Ar^* \quad + \quad Ar$$

Singlet and triplet states are formed with the former often, rather surprisingly, predominating.

II.3 Singlet Oxygen as a Cause of Chemi- and Bioluminescence; the Role of Triplet Oxygen

Many chemiluminescence reactions are oxidation reactions in the direct sense of the word: they require an oxidant such as molecular oxygen, hydrogen peroxide, other peroxides, hypohalites etc. In some cases at least, singlet oxygen is apparently involved in chemiluminescence: in the so-called Trautz-Schorigin reaction (p. 50) and in autoxidation reactions of hydrocarbons, carbonyl derivatives, esters etc. (p. 19). A general chemiluminescence mechanism was proposed [53] with the primary excited species being singlet oxygen – in its monomolecular and its "dimolar" form – the latter being a collisional complex of two O_2 molecules [54]. Chemiluminescence occurs by energy transfer to an appropriate acceptor e. g. a fluorescent reaction product.

This generalization is not corroborated by a sufficient number of experimental facts. Nevertheless however the essential role of singlet oxygen appears to be established in special cases of chemiluminescence: e. g. in the chemiluminescent decomposition of certain aromatic endoperoxides (p. 48). There has been considerable progress in the design of apparatus to measure the characteristically low light levels associated with singlet oxygen [55].

That "normal" triplet oxygen plays an important role in autoxidation chemiluminescence and in many other cases is well known. This will be treated in the appropriate sections of this book.

An unusual *direct* interaction of triplet oxygen has recently been discovered. Strained alkynes react with ground state oxygen (presumably via a dioxet*ene*) as shown [56]:

In this case the conversion of a triplet oxygen collision complex formed with the acetylenic bond to a weakly bound singlet collision complex is discussed, with the possibility that free singlet oxygen may be generated by decomposition of such a collision complex.

In view of the importance of certain biological aspects of singlet oxygen participation e. g. photosensitized oxidation ("photodynamic action"), photocarcinogenity, photochemical smog formation etc., it should be mentioned that an effective method for distinguishing singlet oxygen reactions from those of other oxygen species is regiospecific hydroperoxide formation from cholesterol and some steroids [57, 58]:

The resulting peroxides are different from the products of ground state oxygen interaction and the products can easily be separated by paper chromatography.

C. S. Foote points out that one should not look at a single reaction to distinguish singlet oxygen from other species of oxygen such as $^3O_2$, $O_2^{(-)}$ or $O_2^{2(-)}$ but at a "fingerprint" – a combination of the action of "chemical traps", of quenchers, of the D_2O effect and of chemiluminescence [59].

That compounds known to be quenchers of singlet oxygen can even enhance chemiluminescence from singlet oxygen sources (the hypohalite/hydrogen peroxide reaction, for example) was reported recently [58], 1,4-diazabicyclo-(2,2,2)-octane (DABCO) [60] being one example.

Although there are difficulties in interpretation, it is thought that singlet oxygen is involved in chemiluminescence during autoxidation reactions (p. 19) and in weak bioluminescence (p. 180). However, light emission rather than proof by the methods discussed above seems to be the main criterion for the identification.

There is considerable difficulty in identifying reactive oxygen species in general, and the connections between triplet oxygen, superoxide ion, singlet oxygen and chemiluminescence are particularly confusing. For example, superoxide radical anion $(O_2^{(\pm)})$ is apparently produced from triplet O_2 in solutions of t-butoxide in dimethyl sulphoxide [61]. Yields are not high and if superoxide is required in quantity, electrochemical reduction is of course preferred [62]. The electron affinity of O_2 is 0,43–0,44 eV [63, 64] and it is likely that reducing agents are formed in strongly basic solution. These relationships have been discussed [65]. Since these conditions are often used during the investigation of chemiluminescent compounds such as luminol, the intervention of superoxide must be considered.

Singlet O_2 can be produced from superoxide radical ion by radical ion annihilation (p. 140) using an appropriate oxidant, such as ferrocene [66]. Alternatively, it can be formed in the reaction of KO_2 with diacyl peroxides such as dibenzoyl peroxide or dilauroyl peroxide in benzene in the presence of 18-crown-6 [67].

$$2\ O_2^{(\pm)} + RC\text{-}O\text{-}O\text{-}CR \rightarrow 2\ ^1O_2 + 2\ RCO_2^{(-)} - (R = C_6H_5, C_{11}H_{23})$$

(with $\|$ O below each carbonyl)

Another example in which triplet oxygen takes on the character of another oxygen species is in its reaction with enol ethers [69].

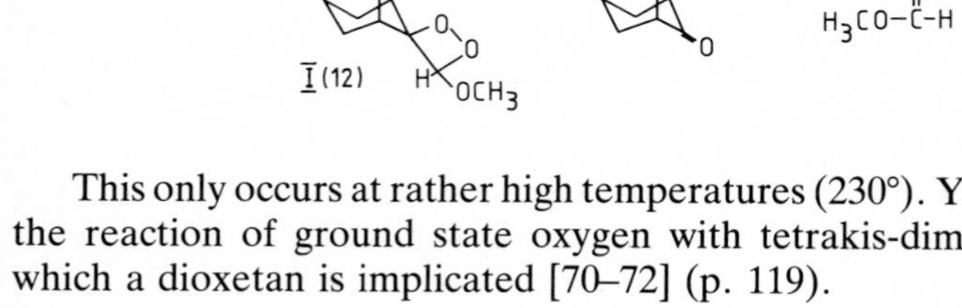

This only occurs at rather high temperatures (230°). Yet another case is that of the reaction of ground state oxygen with tetrakis-dimethylamino-ethylene in which a dioxetan is implicated [70–72] (p. 119).

II.4 References (for Preface, Chaps. I and II)

1. Gundermann, K.-D., Angew. Chem. Int. Ed. Engl. *4*, 566 (1965); McCapra, F., Q, Rev. *20*, 485 (1966)
2. Gundermann, K.-D., Chemilumineszenz organischer Verbindungen, Springer-Verlag Berlin (1968)
3. Radziszewski, B., Ann. *203*, 305 (1880)
4. Radziszewski, B., Chem. Ber. *10*, 70, 321 (1877)
5. Albrecht, H. O., Z. Phys. Chem. *136*, 321 (1928)
6. Gleu, K. and Petsch, P., Angew. Chem. *48*, 57 (1935)
7. Dubois, R., C. R. Soc. biol. *2*, 559 (1885); C. R. Soc. biol. *39*, 564 (1887)
8. Harvey, E. N., Bioluminescence, Academic Press New York (1952)
9. White, E. H., in Light and Life (McElroy, W. D. and Glass, B. eds.) Johns Hopkins Press, Baltimore 1961 p 183; White, E. H. and Bursey, M. M., J. Amer. Chem. Soc. *86*, 941 (1964)
10. Vaughan, W. R., Chem. Rev. *43*, 447 (1948)
11. McCapra, F., J. C. S. Chem. Comm. 155 (1968)
12. Kopecky, K. R. and Mumford, C., Can. J. Chem. *47*, 709 (1969)
13. Hercules, D. M., Acc. Chem. Res. *2*, 301 (1969) – Faulkner, L. R., in Methods in Enzymology, (ed. DeLuca, M.) *57*, 494 (1978)
14. Rauhut, M. M., in Kirk-Othmer, Encyclopedia of Chemical Technology *5*, 416 (1979)
15. Wilson, T. and Hastings, J. W., Photophysics, *5*, 49 (1970)
15 a. Wasserman, H. H. and Murray, R. W. (eds.) Singlet Oxygen, Academic Press, London 1978
16. Proceedings of the International Congress of Chemiluminescence and Bioluminescence: Cormier, M. J. Hercules, D. M., and Lee, J. (Eds.), Plenum Press, New York, London 1973
17. Chemi- and Bio-Energized Processes, Adam, W. and Cilento, G. (Eds) Photochem. Photobiol. *30*, 1–198 (1979)
18. Proceedings of the International Symposium on Analytical Applications of Bioluminescence and Chemiluminescence, Schram, E. and Stanley, P. (Eds.), State Printing and Publishing, Inc. Westlake Village/Calif. 1979
19. Bioluminescence and Chemiluminescence – Basic Chemistry and Analytical Applications, De Luca, M. A. and Mc Elroy, W. D. (Eds.), Academic Press, New York 1981
20 a. Adam, W. and Cilento, G. (eds.) Chemical and Biological Generation of Excited States Academic Press, N. Y. 1982
20 b. Burr, J. (ed.), Chemi- and Bioluminescence, Dekker, N. Y. 1985
21. Hastings, J. W. and Wilson, T., Photochem. Photobiol. *23*, 461 (1976)
22. Schuster, G. B. and Smith, S. P., Adv. Phys. Org. Chem. *18*, 187 (1982)
23. Kamyia, I., Kagaku Kyoiku *28*, 59 (1980)
24. Brandl, H., Naturwiss. Chem. *29*, 65 (1979)
25. Goto, T., J. Syn Org. Chem. *37*, 275 (1979)
26. Gundermann, K.-D., in ref. 18), p. 37
27. Rauhut, M. M., in Kirk-Othmer, Encyclopedia of Chemical Technology, *5*, 419, Wiley, New York 1979
28. Gundermann, K.-D., Top. Curr. Chem. *46*, 61 (1974)
29. White, E. H., Miano, J. D., Watkins, C. J., and Breaux, E. J., Ang. Chem. Int. Ed. Engl. *13*, 299 (1974)
30. McCapra, F., Progr. Org. Chem. *8*, 231 (1973)
31. Rauhut, M. M., Roberts, B. G., Maulding, D. R., Bergmark, W. and Coleman, R., J. Org. Chem. *40*, 330 (1975)

32. Benson, S. W., Thermochemical Kinetics, Wiley, New York 1978
33. Michael, P. R. and Faulkner, L. R., J. Amer. Chem. Soc. *99*, 7754 (1977)
34. Weller, A. and Zachariasse, K. A., Chem. Phys. Lett. *10*, 590 (1971)
35. Schuster, G. B. and Schmidt, S. P., Adv. Phys. Org. Chem. *18*, 187 (1982)
36. Faulkner, L. R. and Bard, A. J., Electroanal. Chem. *10*, 1 (1977)
37. McCapra, F. M., Prog. Org. Chem. *8*, 254 (1973)
38. Kearns, D. R., Chem. Rev. *71*, 395 (1971)
39. Hercules, D. M., Acc. Chem. Res. *2*, 301 (1969)
40. Roberts, D. R., Chem. Commun. 683 (1974)
41. Goddard, W. A. and Harding, L. B., J. Amer. Chem. Soc. *99*, 4520 (1977)
42. Rauhut, M. M., Acc. Chem. Res. *2*, 80 (1969)
43. Calvert, J. G. and Pitts, J. N., Photochemistry Wiley. New York, 1966, p. 88
44. Kasha, M., in Singlet Oxygen (Wasserman, H. H. and Murray, R. W., eds) Academic Press, London 1979 pl.
45. Thrush, B. A., Chem. Br. 287 (1966); NATO Adv. Study Inst. Ser. B 12: 175 (1975); Menzinger, M., Adv. Chem. Phys. *12*, (K. P. Lawley, ed) Wiley, N. Y. 1980, p. 1
46. Wilson, T., Int. Rev. Sci. (2) *9*, 265 (1976); Horn, K. A., Koo, J-Y.; Schmidt, S. P. and Schuster, G. B., Mol. Photochem *9*, 1978–9, 1
47. O'Neal, H. E. and Richardson, W. H. J. Amer. Chem. Soc. *92*, 6553 (1970); Richardson, W. H., Burns, J. H., Price, M. E., Crawford, R., Foster, M., Slusser, P. and Andaregg, J. H., J. Amer. Chem. Soc. *100*, 7596 (1978)
48. Quinga, E. M. Y. and Mendenhall, G. D., ibid. *105*, 6520 (1983)
49. Robinson, G. W. and Digiorgio, V. E., Can. J. Chem. *36*, 31 (1958); Raynes, W. T., J. Chem. Phys. *44*, 2755 (1966)
50. Faulkner, L. R., Int. Rev. Sc. Phys. Chem. Ser. *2*., 9, 213 (1975)
51. Schuster, G. B., Acc. Chem. Res. *12*, 366 (1979)
52. Marcus, R. A., J. Chem. Phys. *52*, 2803 (1970)
53. Khan, A. U. and Kasha, M., J. Amer. Chem. Soc. *88*, 1574 (1966)
54. Spikes, J. D., and Swartz, H. M., Photochem. Photobiol. *28*, 921, esp. 931 (1978)
55. Hurst, J. D. McDonald and Schuster, G. B., J. Amer. Chem. Soc. *104*, 2065 (1982). See there further literature.
56. Turro, N. J., Ramamurthy, V., Liu, K. C., Krebs, A., and Kemper, R. J. Amer. Chem. Soc. *98*, 6758 (1976)
57. Smith, L. L., and Stroud, J. P., Photochem. Photobiol *28*, 479 (1978)
58. Deneke, C. F., and Krinsky, N., ibid. *25*, 299 (1977)
59. Foote, C. S. in lt. 54), p. 922
60. Ouannès, C. and Wilson, T., J. Amer. Chem. Soc. *90*, 6527 (1968)
61. Khan, A. U., Science *168*, 476 (1968)
62. Mayeda, E. A., and Bard, A. J., J. Amer. Chem. Soc. *96*, 4023 (1974)
63. Pack, J. L. and Phelps, A. V., J. Chem. Phys. *44*, 1870 (1966)
64. Celotta, R. J., Bennett, R. A., Hall, J. L., Siegel, M. W., and Lenene, J., Phys. Rev. *6*, 631 (1972)
65. Khan, A. U., Photochem. Photobiol. *28*, 615 (1978)
66. Danen, W. C., and Arudi, R. L., J. Amer. Chem. Soc. *100*, 3944 (1978)
67. Beck, M. T., and Joo, F., Photochem. Photobiol. *16*, 491 (1972)
68. Lechtken, P., Breslow, R., Schmidt, A. H. and Turro, N. J., J. Amer. Chem. Soc. *95*, 3025 (1973)
69. Meijer, F. W. and Wynberg, H., Tetrahedron Lett. *22*, 785 (1981)
70. Paris, J. P., Photochem. Photobiol. *4*, 1059 (1965)
71. Urry, W. H., and Sheeto, J., ibid. *4*, 1067 (1965)
72. Lit. 2, p. 52 ff.

B. Chemiluminescent Reactions

A Profile

We point out at several places in this work, that although there are many luminescent reactions which share common features, there is no single reaction type which provides a unified theory. To assist in ordering the range of reactions with which we deal, the following survey of Chapters III to XI is given.

Autoxidation Reactions

The reactions of Chapter III all involve oxygen, and peroxides are almost invariably intermediates. Many of the mechanisms discovered as a result of the investigations of the isolable peroxides of the succeeding chapter are expected to operate here, but a precise description is often lacking. This represents the more primitive stage of chemiluminescence.

Peroxide Decompositions

More control over the reaction conditions results in isolable peroxides, or at the very least, allows the exact structure of the peroxide to be inferred with some certainty. The reaction mechanisms which form the intellectual foundations of the phenomenon of organic chemiluminescence (and of bioluminescence) are all to be discoved here. In Chapter IV cyclic peroxides display a variety of mechanism, culminating in V with the very important dioxetans. Practical applications of these ideas must not be forgotten, and the chemistry of the active oxalates in Chapter VI brings together previous mechanistic concepts with the most well developed of all the chemiluminescent systems, the active oxalates.

Luminol and Related Compounds

This most familiar of luminescent compounds and its relatives (Chapter VII) also supply several applications. However, in spite of its venerable history, there is no agreed mechanism. The reader must search the earlier descriptions and the literature for clues for a happy ending to this long running story.

Acridine Derivatives

This section (Chapter VIII) starts out with a class of compound which is well understood, and moves on to a discussion of related structures, sharing the common imino-peroxide feature. Mechanisms become less certain however, as we proceed, and there is scope for a variety of explanations.

Miscellaneous Compounds

With one or two exceptions, these reactions (Chapter X) all produce weaker chemiluminescence than the well examined reactions previously introduced. Classification is more difficult.

B. Chemiluminescent Reactions

Electron Transfer Chemiluminescence

This is probably the most basic of the mechanistic ideas and may well underlie the majority of the pathways to luminescence. Electrochemiluminescence gives us an ideal tool with which to study the phenomenon. Closely related chemical systems can also be devised. In contrast to the other chemiluminescent systems, molecular oxygen is not only not required, but must be excluded.

III. Autoxidation Reactions

The majority of these chemiluminescent reactions are weak, but notable exceptions are the oxidation of tretrakis-dimethylamino ethylene (TMAE), p. 119 and the bis-isoquinolinium salts with their reduction product (p. 120). The reaction of O_2 with Grignard reagents is a long known example. Virtually any compound with C-H bonds gives detectable luminescence on oxidation, often with vanishingly low efficiency. It is thus very difficult to derive a convincing mechanism. However the study of simple model compounds has given the chain termination or Russell mechanism [1] a central place in this form of chemiluminescence. In the oxidation of polymers, where the peroxide groups are statistically far removed from each other, it is a less reasonable general mechanism. Recently evidence has been provided for hydrotrioxides as intermediates [54] produced by recombination of HO-radicals with peroxy-radicals (mostly present under the oxidation conditions). Perhaps the gas phase oxidation of methanol [2] to formaldehyde or the second radical annihilation step of the decomposition of hyponitrite esters provide better models.

A particularly effective tool for the study of such reactions was introduced by Vassil'ev [3]. 9,10- Dibromoanthracene, by virtue of the heavy element effect, catalyses acceptance of the triplet excited state energy (residing in carbonyl compounds in this case). Thus the search for the otherwise fairly elusive triplet excited state is made much easier.

III.1 Hydrocarbons

The accepted mechanism of autoxidation is shown below. Initiation can be achieved by the use of readily formed free radicals using for example peroxides of various kinds, or azobisisobutyronitrile. The earliest effective work in this area was undertaken by Vassil'ev and his colleagues [4].

1) Initiation: Formation of radicals from a hydrocarbon $R^1 - CH_2 - R^2$ using initiators such as dibenzoyl peroxide, percarbonates, azobisisobutyronitrile (the radicals yielded by these initiators are generally written, R˙ below)

   $$R^1 - CH_2 - R^2 + R^. \rightarrow R^1 - \overset{.}{C}H- R^2 + R\text{–}H$$

2) Chain Propagation:

   $$R^1 - \overset{.}{C}H - R^2 + O_2 \rightarrow R^1 - CH - R^2$$
   $$|$$
   $$O$$
   $$|$$
   $$O.$$

19

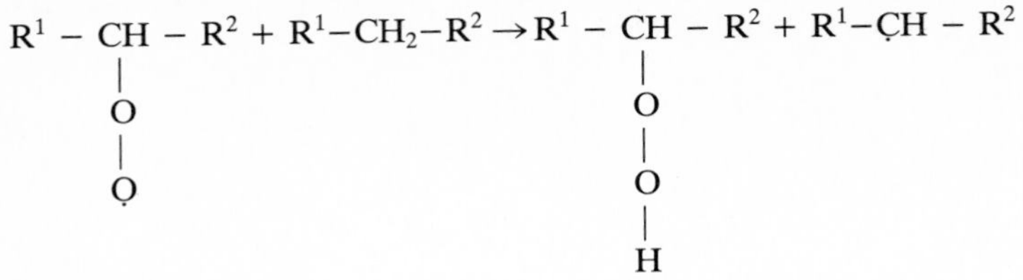

III. Autoxidation Reactions

$$R^1 - CH - R^2 + R^1{-}CH_2{-}R^2 \rightarrow R^1 - CH - R^2 + R^1{-}\overset{\cdot}{C}H - R^2$$

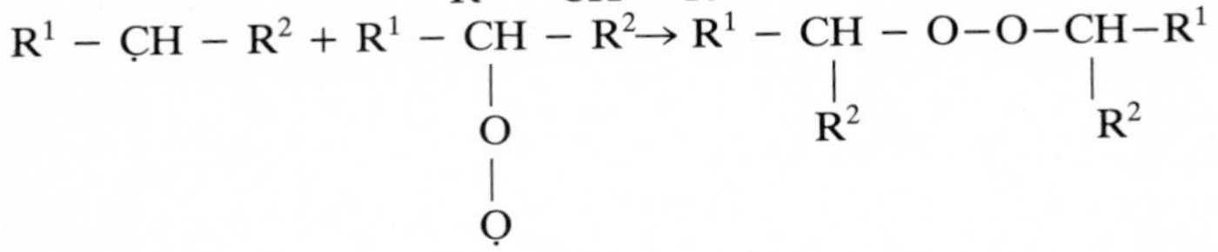

(with the substituents:
$R^1 - CH - R^2$ bearing $-O-\overset{\cdot}{O}$ on the left, and
$R^1 - CH - R^2$ bearing $-O-O-H$ on the right)

3) Chain termination

a) 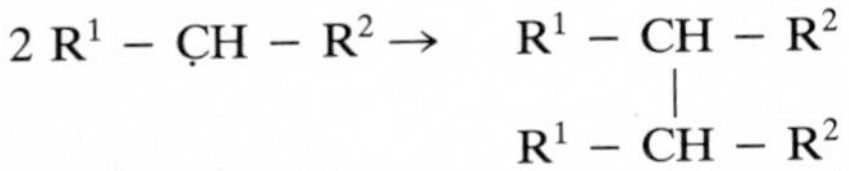 $2\ R^1 - \overset{\cdot}{C}H - R^2 \rightarrow \quad R^1 - CH - R^2$
$$R^1 - CH - R^2$$

b) $R^1 - \overset{\cdot}{C}H - R^2 + R^1 - CH - R^2 \rightarrow R^1 - CH - O-O-CH-R^1$

(with $-O-\overset{\cdot}{O}$ substituent on the left reactant; product bearing R^2 and R^2 substituents)

c) $2\ R^1 - CH - R^2 \rightarrow R^{1\cdot} - CH - R^2 + [R^1 - C - R^2 + O_2]^*$

(left reactant bearing $-O-\overset{\cdot}{O}$; first product bearing $-O-H$; last species bearing $=O$ carbonyl)

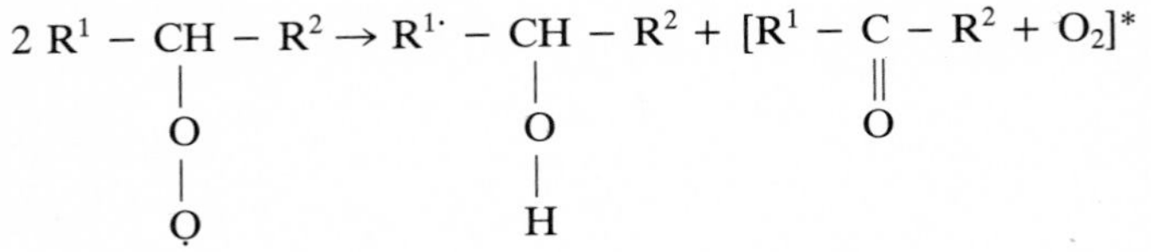

Only the recombination of 2 peroxy radicals 3 c) is, with a ΔH of 483–630 kJ/mol, sufficiently exergonic for the production of excited molecules [5]. Such exothermicity is sufficient to explain the formation of excited states, and the quantum yields are in the range of 0.1 to 40×10^{-9} einsteins/mol. Oxygen quenching is probably responsible for this very low efficiency. The structure of the hydrocarbon and the nature of the cage products also have their effect.

Further quantitative investigations on autoxidation kinetics have led to a more detailed picture of the initiation and of chain termination steps [6, 7]. Details of the steps leading to chemiluminescence are as follows.

$$2\ R{-}O{-}O\cdot \xrightarrow{k_9} R{-}O{-}H + R_2\,C = O + O_2$$

$$2\ R{-}O{-}O\cdot \xrightarrow{k_{10}} P^* (= \text{excited carbonyl compound})$$

$$P^* + Q \xrightarrow{k_{11}} P + Q \text{ (quenching, e.g. by } O_2)$$

$$P^* \xrightarrow{k_{12}} P + h\nu$$

(The rate constants k_1–k_8 not given above pertain to the initiation and chain propagating steps)

If the reaction temperature is held below 50° the concentrations of all chemical species are practically constant over the time (some hours) taken for measurement.

Thus:

reaction rate x $v_{start} = k_{start}$ [initiator] $= v_{termination} = $ const.

light intensity $\dfrac{d(hv)}{dt} = k_{12} [P^*] = \dfrac{k_{10}k_{12}k_{start} \text{ [initiator]}}{k_9 (k_{12} + k_{11} [Q])}$

As $k_{11} [Q] \gg k_{12}$:

$\dfrac{d(hv)}{dt} = k_{12} [P^*] = \dfrac{k_{10}k_{12}k_{start} \text{ [initiator]}}{k_9 k_{11} [Q]}$

Although these reactions are obviously complex good agreement between calculation and experimental observations can be obtained (Fig. 3) for the autoxidation of tetralin in the presence of 10^{-3}M 9,10-diphenylanthracene.

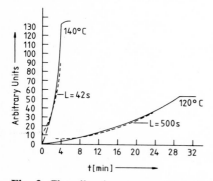

Fig. 3. Chemiluminescence of Tetralin Autoxidation (after Mendenhall and Nathan [6]).

The dotted curves L in Fig. 3 represent the calculated time function related to the time required for the doubling of the chemiluminescence intensity:

$$L = (t_{1/n} - t_{1/n}) \left[\ln \dfrac{n^{1/2} - (1/2)^{1/2}}{n^{1/2} - 1} + 1/2 \ln 2 \right] - 1$$

These calculations assume chain initiation reactions of the following types:

a) $R''OOR'' \rightarrow 2\ R''O\cdot$
    $2\ R''O\cdot + 2 -\!\!\!\underset{/}{\overset{\backslash}{C}} - H \rightarrow 2\ R''OH + 2 -\!\!\!\underset{/}{\overset{\backslash}{C}}\cdot$
b) $R''OOH + $ catalyst $\rightarrow R''O\cdot$
    $R''O\cdot + -\!\!\!\underset{/}{\overset{\backslash}{C}}-H \rightarrow R''OH + -\!\!\!\underset{/}{\overset{\backslash}{C}}\cdot$
c) $-\!\!\!\underset{/}{\overset{\backslash}{C}}-H + O_2 \rightarrow HOO\cdot + -\!\!\!\underset{}{\overset{\backslash}{C}}\cdot$

Apart from the complexity of such calculations, there are often additional problems such as the formation of phenolic inhibitors during the autoxidation reaction, possibly causing a considerable decrease in oxygen uptake [8, 9]. These initiation and propagating steps are not directly responsible for the light emission, but provide the peroxides whose decomposition gives rise to the excited states.

III.1.1 The Excitation Step

As previously mentioned autoxidation chemiluminescence is invariably of low quantum efficiency. The range observed is shown in Table 1.

Table 1. Chemiluminescence quantum yields of some autoxidation reactions [10]

Compound	Chemiluminescence quantum yield (Einstein/mole) $\times 10^9$
Cyclopentanone	40
Benzylphenylketone	8
2-Butanone	7
2-Heptanone	2.4
Cyclohexane	2.4
Cyclododecane	1.1
Ethylbenzene	0.9
n-Heptane	0.8
n-Octane	0.16
n-Dodecane	0.15

Initiator: dicyclohexyl peroxydicarbonate DCPD), 10^{-2} mole. All compounds oxidized without solvent except benzylphenylketone which was dissolved in chlorobenzene.

The quantum yields were calculated on the following assumptions: A molecule of initiator (DCPD) produces 2 radicals on homolysis. Each radical, in turn, yields one molecule of peroxide, which decomposes in a bimolecular reaction to ketone and alcohol.

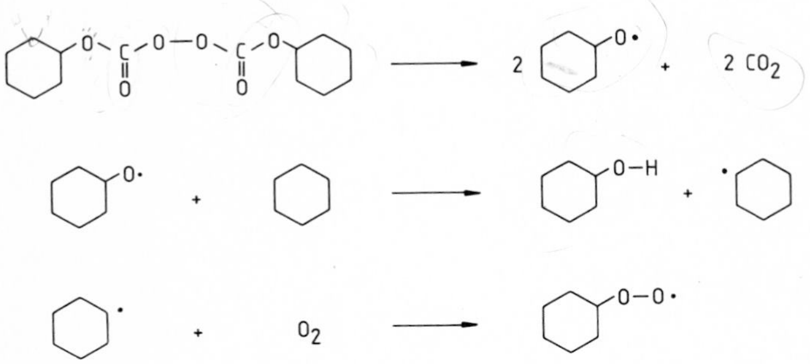

The termination step occurs by the Russell mechanism [1] via the unstable tetraoxide (5). The products are a ketone, oxygen and an alcohol as shown. The conservation of spin in this assumed concerted reaction, as previously discussed, leads, to formation of the ketone in the triplet excited state. Although singlet oxygen may be found in low yield, the carbonyl excited state certainly predominates.

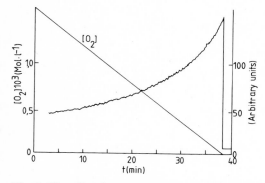

The claim [12] that the spectrum of 2-butanone chemiluminescence is identical with that of product (biacetyl) phosphorescence, has been confirmed [10].

This mechanism also helps explain the very low quantum yields. Strong quenching of the excited carbonyl triplets by the (triplet) oxygen simultaneously formed in the solvent cage is very likely. This quenching is so strong that only one in 10^8 excited molecules is likely to emit. The rate of triplet-triplet transfer quenching was estimated to be about $10^{11}s^{-1}$. The *excitation yield* might then be nearly unity in some cases although the highest quantum yields of autoxidation chemiluminescence measured were 4.0×10^{-8}.

Oxygen quenching is also responsible for the remarkable kinetic behaviour of the chemiluminescent oxidation of hydrocarbons in closed vessels. Instead of the expected gradual decrease of intensity as oxygen is consumed, the intensity actually rises (Fig. 4) Depletion of O_2 results in such an enhancement of emission that at the very point at which light emission should be negligible, it is maximal. The cessation of oxidation which inevitably results from consumption is all the more dramatic – the "oxygen drop" phenomenon [11].

Small quantities of oxygen can be measured using this effect (p. 176).

Fig. 4. Chemiluminescence intensity vs. time plot of ethylbenzene autoxidation in benzene at 40°. Initiator: dicyclohexylperoxycarbonate ($5,2 \times 10^{-2}$ M). The straight line was drawn assuming an initial oxygen concentration calculated from solubility data in the literature and assuming that at the "drop", oxygen concentration equals zero [12]).

Although differences in structure give rise to differences in quantum yield (Table 1) several factors are involved and no clear pattern can be seen. Some doubt is cast on the conclusion (see p. 23) that after allowing for the effects of oxygen quenching, unit quantum yields are expected. In the case of dimedone oxidation the excited product, an α-diketone, is known not to be susceptible to O_2 quenching [13], yet the quantum yield is only 10^{-8}.

$\underline{\underline{\text{III}}}$ 1 $\qquad$ $\underline{\underline{\text{III}}}$ 2 $\qquad$ $\underline{\underline{\text{III}}}$ 3

Other mechanisms may exist for quenching, although it is difficult to estimate the effect of the formation of oxygen within the cage. It seems that phosphorescence from the excited carbonyl products in these reactions is always accompanied by singlet oxygen emission [3,14]. It is not possible to separate the direct formation of $^1O_2$ (as an alternative result of the spin conservation previously mentioned) from energy transfer quenching of the triplet carbonyl.

In this case Kellogg [10] observed emission in the singlet oxygen range of the spectrum. It was exceedingly weak ($\emptyset = 0.9 \times 10^{-11}$ and it was not possible to ascribe it unambiguously to the emission from $^1O_2$.

Spectroscopic evidence for the generation of singlet oxygen was apparently found in the recombination of sec-peroxy radicals. Thus the emission spectrum of linoleic acid autoxidation chemiluminescence showed 5 characteristic peaks around 470–480, 510–530, 560–580, 630–640 and 670 nm, in good agreement with the maxima of the chemiluminescence exhibited during sec-hydroperoxide oxidation by Ce^{4+}-ions [14]. The maxima were ascribed to the ($^1\Sigma_g^+$, $^1\Delta g$) and ($^1\Delta g$, $^1\Delta g$) dimers of singlet oxygen [15, 16].

This conclusion has been challenged [17] as certain reactions, occurring in the peroxy/ Ce^{4+} system, were observed also in the absence of peroxy radicals.

Other compounds whose autoxidation and chemiluminescence have been studied are tetralin [18–20], cumene and ethylbenzene [20,21]. Oscillating chemiluminescence during such oxidations has also been observed [22].

The complexity of these radical chain processes precludes a detailed examination of the formation of the excited state in the manner of other chemiluminescent compounds.

III.1.2 Polymers

This is a field of great practical interest since it provides a means of studying the stability of polymers, e.g. polystyrene, polymethyl methacrylate, polycarbonates, polypropylene etc. towards oxidative degradation.

Initiation during such degradation is often the result of irradiation by ambient light, but an alternative initiator used in the study of polymer chemiluminescence is dicyclohexyl- peroxydicarbonate (DCPD) [23].

The chemiluminescence emission in polystyrene and in polycarbonate has maxima at 450 and 530 nm which reach their highest intensity ca. 60 s after the start of the thermolysis at 70–80°, with DCPD as initiator.

The 530 nm maximum was ascribed to triplet cyclohexanone produced in the familiar chain termination step:

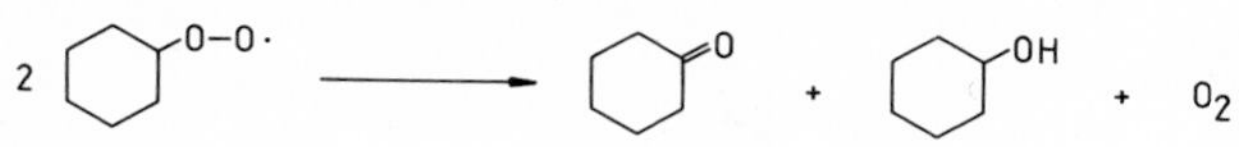

III (4)

The peroxy radical is formed from cyclohexyloxy radicals stemming from the initiator DCPD via the following reaction steps [3] as described on p. 22. It is possible to relate the initial burst of chemiluminescence from polymeric materials to the pre-existing peroxide concentration, and to obtain an estimate of their useful life in air.

As the emission is very weak, energy transfer to 9,10 – dibromo anthracene (DBA) and other compounds (biacetyl, bibenzyl, naphthalene or anthracene was used [24,24 A].

If DCPD also initiates an autoxidation radical chain in the polymers the peroxy radical grouping should be formed at tertiary and at secondary C-atoms:

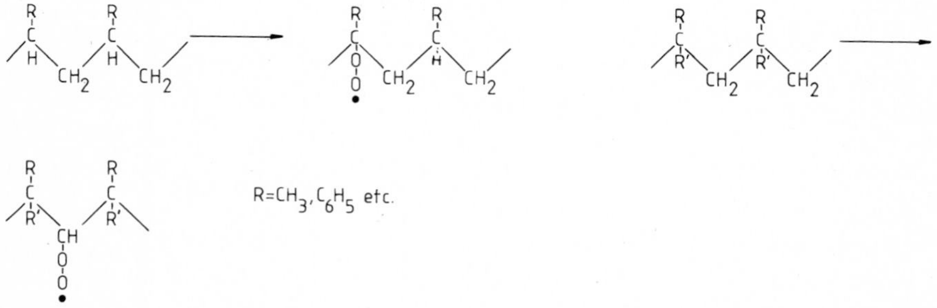

$R = CH_3, C_6H_5$ etc.

Ketones could be formed from the secondary hydroperoxides. It has been reported [25] that oxidation of polypropylene yields ketones, esters, carboxylic acids, and alcohols.

A tertiary peroxy radical should finally lead to cleavage of the polymer chain:

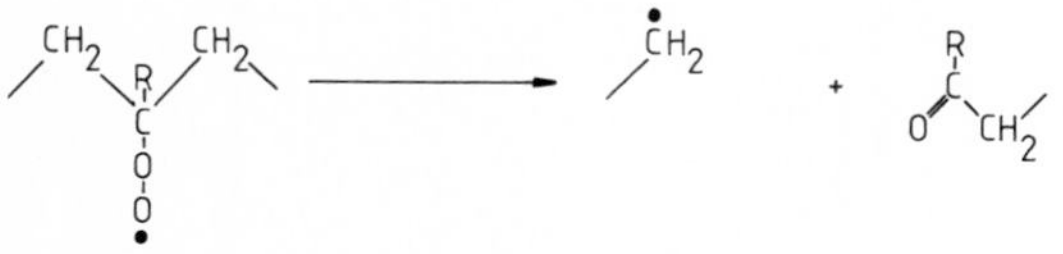

However, it is very improbable that a radical chain reaction can proceed in relatively immobile polymers. Thus the chemiluminescence observed as well as the previously mentioned carbonyl groups may not be caused by the Russell mechanism (perhaps via trioxides [54]).

Yet as the effect of an initiator can be inhibited by appropriate compounds (see p. 26, 30) indirect but practically useful conclusions can be drawn concerning the stabilization of macromolecular materials against the combined effects of oxygen (air) and radiation, especially ultraviolet light. As an example stable iminoxy radicals of the type 5 were studied as to their inhibitory effect on the oxidative degradation of polyethylene oxide and polystyrene [26].

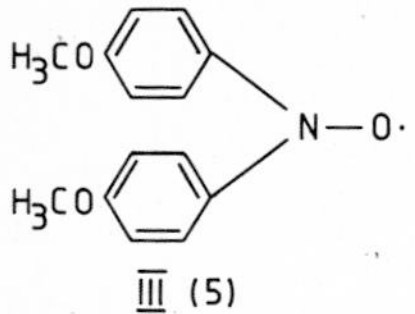

III (5)

III.1.3 Base-catalysed Autoxidation of Carbonyl Compounds

The simplest compounds which emit light on autoxidation in basic solution are ketones and esters. Although the reactions of alkyl substituted compounds are never efficient ($\emptyset_{CL}$ 10^{-4}–10^{-8}), the mechanism of reaction has much in common with some of the most efficient of all compounds such as the luciferins and acridan esters.

The expected intermediate in all cases is the α-hydroperoxide, which in many instances is isolable. Richardson [27] investigated the decomposition of (6) and (7) in basic methanol

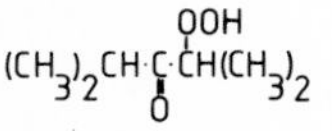

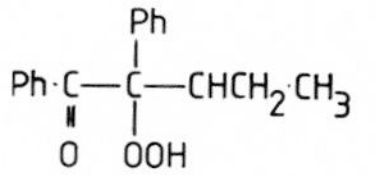

III (6)   III (7)

Even in anhydrous MeOH with NaOMe as catalyst butyric acid (in the case of (6)) was formed in 70% yield. Thus attack on the carbonyl group must be by the peroxide anion (presumably to form a transient dioxetan) rather than the external nucleophile methoxide which would yield the ester (some methyl isobutyrate is formed in very low yield). The reaction scheme is shown below. Blue-green light is visible from the reaction.

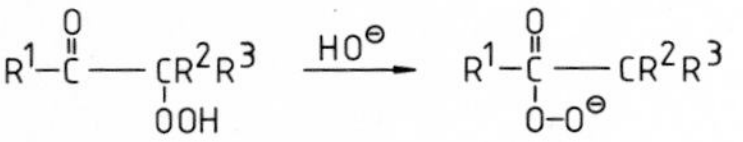

III (8)

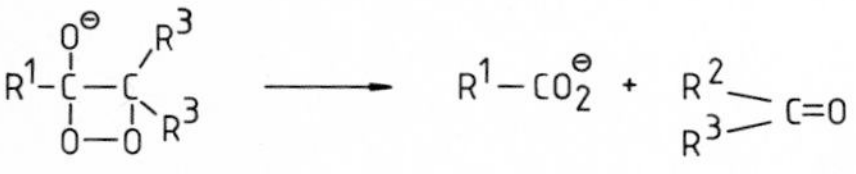

This simple picture does not obtain if the conditions are changed. For example [28] in the autoxidation of the indanone (9) the ester is formed, and the

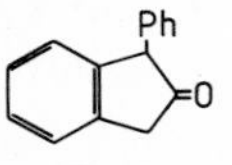

III (9)

26

authors assume that the dioxetan route is discriminated against by the extra ring strain produced by the five-membered ring.

Ogata and Sawaki [29] examined a large number of α-hydroperoxyketones of the general formula (8) and found, in benzene-methanol, that methyl esters were obtainable in yields up to 100%. Although the compound studied by Richardson gave low yields of ester (about 30%), there is clearly a difference in behaviour. Quantum yields (singlet: triplet = 1: 300) were low ($\emptyset_{CL} = 10^{-6}$). The conclusion is that dioxetan formation occurs to a small extent, and is responsible for the light emission. Other similar studies [29, 30] show that the route taken depends on several factors, and that hindered bases such as potassium *tert*-butoxide are more likely to result in the cyclic (dioxetan) route with formation of the acid rather than ester.

Base catalysed oxidation of simple aryl carboxylic acid esters has been studied [31, 32[and light is emitted in the presence of fluorescers [33]. The quantum yield is low ($\emptyset_{CL} = 10^{-6}$) and an electron transfer chemiluminescence seems likely in view of the order of efficacy of the fluorescers, fluorescein $> 9,10$-diphenyl anthracene $>> 9,10$-dibromoanthracene.

Much more light is obtainable [34] from the autoxidation of anthracenyl ketones such as (10)

Yellow light was emitted, corresponding to the fluorescence of the acid product. No emission was observed from the monoacyl anthracene, although the carboxylic acid product is fluorescent. Acylcarbazoles behave similarly [35]. A fairly extensive study has been made of 1,8- and 1,5-diacyl anthracenes [36]. It was shown that primary and secondary (11, R_1 or $R_2 = H$) ketones gave less light than tertiary ketones ($R_1 = R_2 = $ alkyl). This can be explained by base catalysed elimination of water from the intermediate peroxide when an α-H atom ist present. The chemiluminescence spectra did not match the spectra of fluorescent products, being considerably red-shifted. Complex formation (e. g. an exciplex) was suspected.

The value of electron donation in any oxidative chemiluminescence is seen in the chemiluminescence of (12) under autoxidative conditions. The quantum yield is higher than well known bright compounds such as luminol, $\emptyset$ being 4.6%!

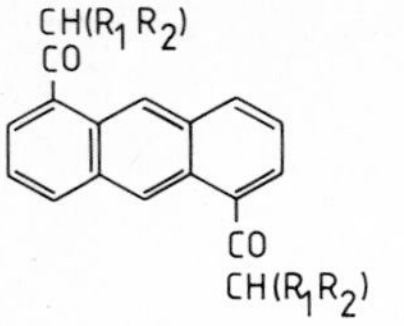

III (11)

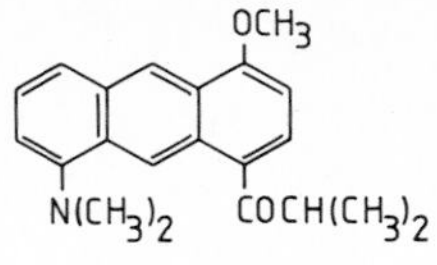

III (12)

The autoxidation of acridan esters has been of great value in the elucidation of the mechanism of luciferin oxidation, and is discussed as a model system for bioluminescence (Chap. (XII)). The following series of compounds which react in a related way summarises the main features which lead to moderately high light emission from the autoxidation of carbonyl compounds. The alkoxycarbamoyl furanones [37], whose chemiluminescent reactions are depicted in the scheme below, show the following useful features.

(1) An acidic α-H atom whose removal (by O_2) leaves a specially stabilised carbon radical.
(2) A good leaving group which is expelled on attack by the peroxide to give a dioxetan (or in this case dioxetanone).
(3) A fluorescent product of low ionisation potential.

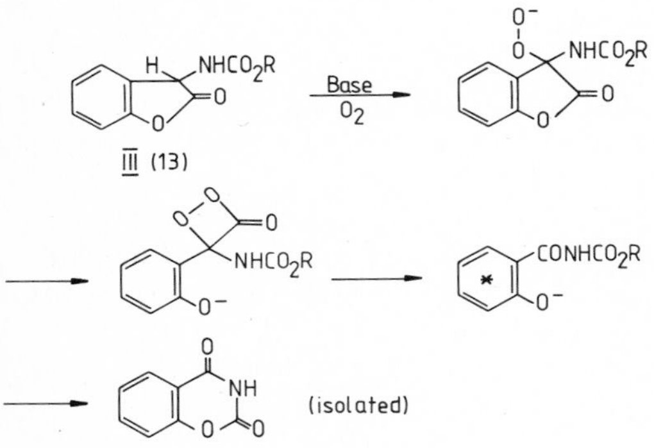

III (13)

(isolated)

This is the pattern of the luciferins and many related compounds and is one of the more secure mechanisms for chemiluminescence.

III.1.4 Grignard Compounds [38]

The chemiluminescence of the reaction of Grignard reagents with oxygen was noted as early as 1906 [39], but only recently has some progress been made in understanding the mechanism of this reaction [40, 41]. The following facts appear to be clearly established:

1) Free radicals are formed during the reaction (proved by the behaviour of the reaction mixture in an external magnetic field, and by ESR).
2) The emission has maxima at 330 nm [41] or 357 nm [40], making Grignard chemiluminescence one of the most exergonic reactions of this type, requiring reaction enthalpies of at least 378 kJ/mol.

There has been disagreement about the emitting species. *p-Terphenyl* (formed in 0.1% yield) was suggested [40] as the emitter since the emission spectrum of phenyl magnesium bromide oxidation matches the fluorescence spectrum of terphenyl obtained from an "authentic sample" (40).

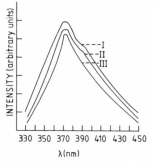

Fig. 5. Fluorescence spectrum of oxygenation products of PhMgBr in diethyl ether; wavelenght of excitation: 313 nm *(I)*. Fluorescence spectrum of photolyzed bromobenzene in diethyl ether; wavelength of excitation: 313 nm *(II)*. Chemiluminescence spectrum of the reaction product(s) of a diethyl ether solution of PhMgBr with oxygen *(III)* (after [41]).

This statement, however, has been challenged [41] because "little resemblance of that spectrum to the literature spectrum" [42] was found. An isomeric mixture of brominated biphenyls and an associated reaction mechanism were proposed instead [41]. It has been pointed out [43] that authentic brominated biphenyls have not been available for comparison.

Figure 5 shows the fluorescence spectrum of the oxidation products of phenyl magnesium bromide in diethyl ether and the chemiluminescence spectrum of the oxidation of phenyl magnesium bromide in the same solvent. The fluorescence spectrum of photolyzed bromobenzene is also shown for comparison.

To account for the high energy produced by the reaction of phenyl magnesium bromide and related compounds (PhMgX) with oxygen the following steps are proposed [41].

$$PhMgX + O_2 \rightarrow PhO_2MgX \rightarrow Ph\cdot + MgO_2X$$
$$Ph\cdot + MgX_2 \rightarrow PhMgX + X\cdot \qquad \text{(Formation of halogen radicals)}$$
$$X\cdot + PhMgX \xrightarrow{R\cdot} XPhMgX + RH \qquad \text{(Formation of halogenated phenyl group)}$$

$$X\,Ph\,MgX + O_2 \rightarrow XPhO_2MgX$$
$$XPhO_2MgX + PhMgX \rightarrow XPh–Ph^* + 2\,MgOX \qquad \text{(Formation of excited biphenyl}$$
$$XPh–Ph^* \rightarrow XPh–Ph + h\nu \qquad\qquad\qquad \text{derivative)}$$
$$XPhO_2MgX + PhMgX \rightarrow XPhMgX + PhOMgX$$

An overall exothermicity of at least 420 kJ is calculated, the highest contribution assumed to be from the formation of the halogenated biphenyl by the reaction:

$$XPh\cdot \text{ (from } XPhO_2MgX) + Ph\cdot \text{ (from starting Grignard compound)} \rightarrow XPh–Ph$$
$$(-\Delta H \sim 588\,kJ/mole)$$

29

It is assumed [40, 41] that the emitting species is not formed by energy transfer from some excited primary product but directly from a peroxide intermediate.

III.1.5 Inhibitors

The principal action of inhibitors in autoxidation chemiluminescence probably lies in the trapping of the peroxy radicals [44]:

$$RO_2 \cdot + \text{ inhibitor} \rightarrow \text{stable products}$$

Thus the decrease in light yield is most often an inhibition rather than quenching.

Stable iminoxyl radicals of the structure type $Ar_2N - O \cdot$ [45, 46], substituted phenols [47, 48], and metal-, especially zinc -dialkyl dithio-phosphates [49] are the most common inhibitors used.

III.1.6 Miscellaneous Oxidation Reactions

III.1.6.1 Oxidative Chemiluminescence of Coals

Treatment of hard coal with oxygen at temperatures of ca. 100° results in chemiluminescence, the intensity of which is dependent on the dispersion of the coal, and on previous contact with air. The origin of the coal affects the chemiluminescence [50]. Similar observations have been made on the pit waste, no doubt as a result of coal inclusions [51].

No detailed mechanistic investigations have been reported, but it is reasonable to assume a close relationship with the previously discussed hydrocarbon autoxidations. These Russian reports have not been pursued in other regions, but the result may have application in the comparison of coal deposits.

III.1.6.2 Cysteine and Glutathione

Chemiluminescence has been observed during the oxidation of these compounds by molecular oxygen in the presence of heavy metal catalysts. Copper-II-compounds ($[Cu(NH_3)_4]^{2(+)}$ and a Cu- flavin mononucleotide complex) were found to be especially efficient. The emitting species seems to be singlet oxygen in both cases. The emission is accompanied by FMN fluorescence [52].

The singlet oxygen may be formed from the superoxide anion $O_2^{(\pm)}$ produced by a one-electron transfer from the sulfur atoms of cysteine or glutathione, respectively. The products are the corresponding disulfides. The one- electron transfer appears to take place in the coordination sphere of e.g. the copper-II ions, according to spectroscopic evidence.

$(Cu\,X_n)^{2(+)} + O_2 \rightarrow (CuX_nO_2)^{2(+)}$ ($X = NH_3$, FMN)
$(Cu\,X_nO_2)^{2(+)} + $ cysteine anion $\rightarrow (CuX_{n-1}\text{cysteine})^{1(+)} + X$
$(Cu\,X_{n-1}\text{ cysteine} - O_2)^{1(+)} \rightleftharpoons (CuX_{n-1}\text{cysteine})^{2(+)} + O_2^{(\pm)}$
$(Cu\,X_{n-1}\text{ cysteine} - O_2)^{1(+)} + $ cysteine $+ X$

$$\rightarrow (Cu\,X_n - O_2H)^{1(+)} + \text{cysteine}$$
$(Cu\,X_n - O_2H)^{1(+)} \qquad\qquad \rightleftharpoons (Cu\,X_n)^{2(+)} + O_2H$

Another chemiluminescent system containing copper-II- ions, riboflavin, hydrogen peroxide, and mercaptoethanol [53] appears to proceed via a similar mechanism, $HO_2\cdot$ radicals being intermediates.

III.2 References

1. Russell, G. A., J. Amer. Chem. Soc. *79*, 3871 (1957)
2a. Vaidya, W. M., Proc. Roy. Soc. A *219*, 572 (1964)
2b. Menzinger, M., Adv. Chem. Phys. *12*, 1 (1980)
3. Belyakov, V. A. and Vassil'ev, R. F., Photochem. Photobiol. *11*, 179 (1970)
4. Vasil'ev, R. F. and Belyakov, V. A., Photochem. Photobiol. *6*, 35 (1967) and references cited
5. Vasil'ev, R. F., Progr. React. Kin. *4*, 305 (1967)
6. Mendenhall, D. G., and Nathan, R. A., Angew. Chem. *89*, 220 (1977); Intern. Ed. Engl. *16*, 225 (1977)
7. Lundeen, G. and Livingston, R., Photochem. Photobiol. *4*, 1085 (1965)
8. Robertson, A. and Waters, W. A., J. Chem. Soc. *1948*, 1574
9. Woodward, A. E., and Mesrobian, R. B., J. Amer. Chem. Soc. *75*, 6189 (1953)
10. Kellogg, H. J. Amer. Chem. Soc. *91*, 5433 (1969)
11. Vasil'ev, R. F. and Rusina, I., Doklad. Akad. SSSR *153*, 1101 (1963)
12. Vasil'ev, R. F. and Rusina,, I. ibid. *153*, 5433 (1963)
13. Beutel, J., J. Amer. Chem. Soc. *93*, 2615 (1971)
14. Howard, J. A. and Ingold, K. U., ibid. *90*, 1056 (1968)
15. Nakano, M., Takayama, K. and Shimizu, Y., ibid. *98*, 1974 (1976)
16. Murray, R. W., in Lit. 78), p. 59, 84
17. Krinsky, N. I., in Lit. 78) p. 597, esp. 603
18. Howard, J. A., Schwalin, W. J. and Ingold, K. U., Advan. Chemistry Series No. *75*, 6 (1968)
19. Howard, J. A., Ingold, K. U. and Symonds, M., Can. Chem. J. *46*, 1017 (1968)
20. Middleton, B. S. and Ingold, K. U., Can. J. Chem. *45*, 191 (1967)
21. Lazar, M. and Matisová-Rychla, L., J. Luminescence *6*, 167 (1973)
22. Rychlý, J., Matisová-Rychlá, L. and Lazar, M., J. Polymer. Sci. Sym. *57*, 139 (1976)
23. Philipps, D., Anissimov, V., Karpukhin, O. and Schlyapintokh, V. Ya., Photochem. Photobiol. *9*, 183 (1969)
24. Vasil'ev, R. F. and Belyakov, V. A., ibid. *11*, 179 (1970)
24a. Förster, T., Discuss. Farad. Soc. *27*, 7 (1959)
25. De Kock, R. J. and P. A. H., M. Hol, Rec. Trav. chim. Pays-Bas *85*, 102 (1966)
26. Matisová-Rychlá, L., Lazar, M., Rychlý, J. and Karpukhin, O. N., J. Polym. Sci. Sym. *40*, 145 (1973)
27. Richardson, W. H., Hodge, V. F., Stiggall, D. L., Yelvington, M. B. and Montgomery, F. C., J. Amer. Chem. Soc. *96*, 6652 (1973)
28. Bordwell, F. B. and Knight, A. C., ibid. *93*, 3416 (1971)
29. Ogata, Y. and Sawaki, Y., ibid. *97*, 6983 (1975)
30. Ogata, Y. and Sawaki, Y., ibid. *99*, 5412 (1977)
31. Kamiya, I. and Sugimoto, T., Bull. Chem. Soc. Japan *50*, 2442 (1977)
32. Avramoff, M. and Sprinzak, Y., J. Amer. Chem. Soc. *85*, 1655 (1963)
33. Ogata, Y. and Sawaki, Y., J. Org. Chem. *42* 40 (1977)
34. Hiramatsu, T., Harada, T. and Yamaji, T., Bull. Chem. Soc. Japan *55*, 985 (1982)
35. Kamiya, I. and Sugimoto, T., ibid. *54*, 25 (1981)

36. Kamiya, I. and Sugimoto, T., Chem. Lett., 335 (1978); Photochem. Photobiol. *30*, 49 (1979)
37. Lofthouse, G., Suschitzky, H., Wakefield, B., Whitaker, R. and Tuck, B., J. Chem. Soc., Perkin Trans 1, 1634 (1979)
38. See Gundermann, K.-D., Topics Curr. Chem. *46*, 63 (1974), p. 75
39. Wedekind, F., Z. wiss. Phot. *5*, 29 (1905)
40. Bardsley, R. L. and Hercules, D. M., J. Amer. Chem. Soc. *90*, 4545 (1968)
41. Bolton, P. H. and Kearns, D. R., ibid. *96*, 4651 (1974)
42. Berlman, I. B., Handbook of Fluorescence Spectra of Aromatic Compounds Academic Press N. Y. 1971
43. Hastings, J. W. and Wilson, T., Photochem. Photobiol. *23*, 461 (1976)
44. Shlyapintokh, V. Ya., Chemiluminescentnye Metody Issledovania Medlennych Chimiceskich Processov, p. 158, 1960
45. Thomas, J. R., J. Amer. Chem. Soc. *82*, 5955 (1960)
46. Matisová-Rychlá, L., Karpukhin, O. N.and Pochalok, T. O., Photochem. Photobiol. *18*, 303 (1973)
47. Orudsheva, I. M., Suleimanova, L. G. and Liksha, V. B., Azerb. Khim. Zhurn. *1974*, 51; C. A. *82*, 139 × (1975)
48. Matisová-Rychlá, L., Ambrovic, P., Kulicková, N. and Rychlý, J., J. Polym. Sci., Symposium No. *57*, 181 (1976). This paper describes the influence of 2,4,6-trisubstituted phenols (e.g. 2,6-di (t-butyl)-4-alkyl- or -cycloalkyl phenols on the autoxidation chemiluminescence of polypropylene
49. Ivanov, S. K., Yuritsin, V. S. and Shopov, D., Coll.Czech. Chem. Commun. *37*, 3284 (1972); C. A. *79*, 12 5595 k (1973)
50. Kucher, R. V., I. A. Opeida and I. N. Dumbai, Khimiya tv. topliva (Solid Fuel Chemistry) *9*, (1975), Nr. 4
51. Dumbai, I. N., Kucher, V. R. and Sarancuk, V. J., ibid. *13*, 67 (1979)
52. Stauff, J. and Nimmerfall, F., Z. Naturforsch. *25 b*, 1009 (1969)
53. Steele, R. H. and Vorhaben, J. E., Biochemistry *6*, 1404 (1967)
54. Shereshovets, V. V. et al., Izv. Akad. Nauk SSR Ser. Khim. *1982*, 2631

IV. Chemiluminescent Peroxide Decompositions, I (except Dioxetans)

The great majority of chemiluminescent reactions in solution require molecular oxygen. It is to be expected that peroxides are almost universal intermediates. Study of the luminescent reactions of peroxides of known structure is clearly an important part of the investigation.

The mechanisms fall into two broad classes. In the first there are structural features which predispose the reaction towards the formation of excited states. Two important divisions of this class are represented by peroxides undergoing reaction by chemically initiated electron exchange luminescence (CIEEL) (p. 34) and the dioxetans. However, many of the luminescent reactions of peroxides are not dependent on a unique structure for the peroxide and the exact mechanism of excitation can be as difficult to elucidate as in the cases where the peroxides are not isolated. Dioxetans are a special case and are dealt with in detail in Chap. V.

IV.1 Peroxide Chemiluminescence by CIEEL

The study of electrogenerated chemiluminescence is inherently more precise than that of the peroxides. It is thus fairly certain that the electron transfer described by Marcus [1] is a sound foundation on which to build. The excitation step involves the donation of an electron from the radical anion formed by reduction at the *cathode* to the radical cation formed by oxidation at the *anode*. Given certain structural requirements, it appears that peroxides can serve as the oxidant.

It is of crucial importance that the transfer of the electron back to the radical cation is facilitated by the structure of the peroxide reduction products. This idea was first indicated by McCapra [2] in an explanation of the very efficient active oxalate chemiluminescence, and serves as a good example of the above mechanism.

We assume, for the present, that the intermediate peroxide is the dioxetan-dione (13), although evidence for its involvement is discussed elsewhere (p. 75).

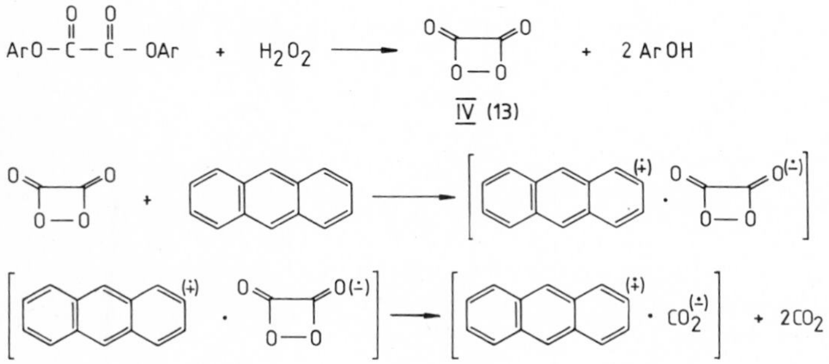

At the time of this proposal, it had been made clear by the work of Rauhut and his coworkers (see p. 70) that the ionisation potential of the aromatic hydrocarbon (anthracene in the example above) had to be low for high light yields. This can be explained by the need for the reduction of the very unstable cyclic peroxide to have a large rate constant compared to those for dark reactions. The essential feature is the way in which fragmentation of the peroxide on acquisition of an electron can result in stable molecules (CO_2) on return of the electron to the radical cation.

The term CIEEL was coined by Schuster [3] in a re-appraisal of the chemiluminescence of diphenoyl peroxide [4] (14).

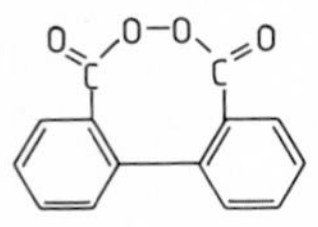

$\underline{IV}$ (14)

Extensive work has been done to demonstrate its generality.

Essentially the same idea had been advanced by Linschitz [5] as early as 1961 to explain the fact that tertiary hydroperoxides were relatively poor in comparison with secondary peroxides in eliciting chemiluminescence from metallo-porphyrins such as zinc tetraphenylporphin (ZnTPP). It was suggested that the peroxide was reduced by the transfer of an electron from the metallo-porphyrin forming an alkoxy-radical and the charge transfer complex ZnTPP$^{(+)}$HO$^{(-)}$. The reaction scheme is shown using the most convenient peroxide, tetralin hydroperoxide.

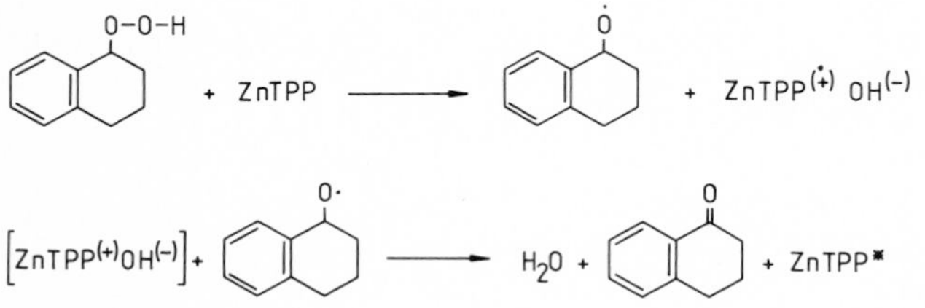

Excitation of the metallo-porphyrin occurs by transfer of an electron to the ZnTPP radical cation, yielding α-tetralone and water as the other products. The reaction of ZnTPP with other peroxides known to generate excited carbonyl products by other mechanisms has also been examined [8]. Chemiluminescence is inefficient in all these other cases confirming the special features claimed for *direct* formation of excited states from secondary peroxides.

The importance of a charge transfer complex between the peroxide and the metal and the value of a low ionisation potential for the metalloporphyrin has been demonstrated using dimethyl dioxetanone [6]. Typical CIEEL behaviour was observed, and evidence for the importance of a charge transfer complex between the peroxide and the metallo-porphyrin obtained from competitive complexation experiments using pyridine and diethyl ether. Considerable reductions in the rate of reaction were observed.

34

IV.2 Acyclic sec.-Peroxy-esters

The mechanism of the chemiluminescent decomposition of secondary peroxy-esters shows similar features [7], in that the loss of the secondary proton allows electron transfer to the activator radical cation. Activators are fluorescent compounds of low ionisation potential. In the example shown below the activator is not consumed, and the products, acetophenone and acetic acid, are formed quantitatively. The unactivated decomposition is very closely related to peroxide decomposition generally, and it is not surprising to find that the quantum yield is very low.

Addition of a fluorescer, such as the easily oxidised dihydro-dibenzo-[a, c]-phenazine gives a much enhanced light yield and the reaction then has much in common with the mechanism proposed by Linschitz for ZnTPP activated luminescence reactions previously discussed.

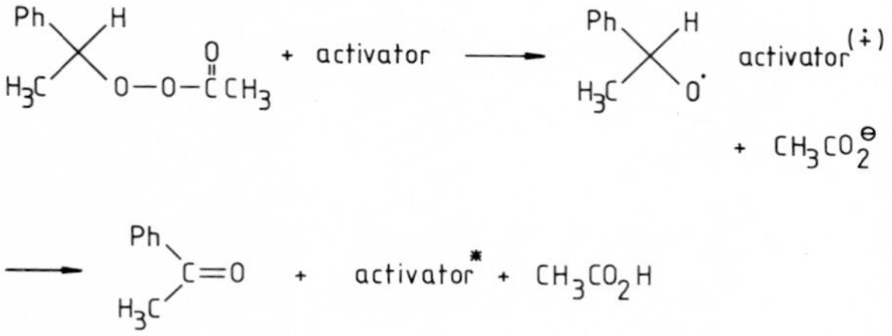

The cleavage of the peroxide bond is thought to be preceded by electron transfer from the fluorescer and then back transfer occurs from the ketyl formed by loss of a proton from the secondary peroxide.

Some models of the excitation reaction in bacterial bioluminescence seem to require a similar explanation [8].

Biochemical studies [9] have resulted in the scheme shown:

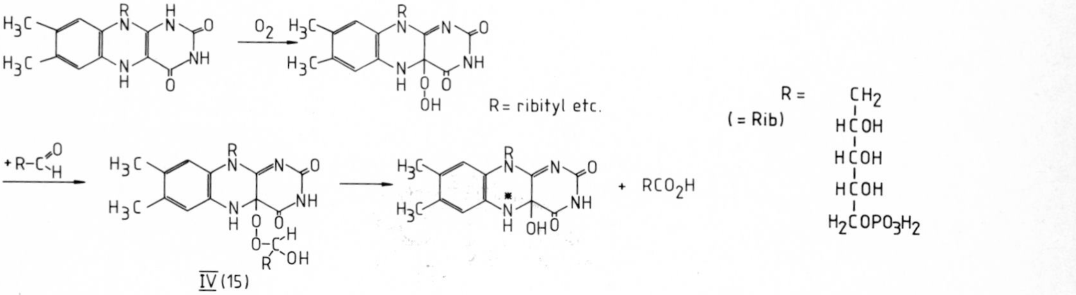

There is no direct evidence for the intermediate (15) but its formation seems inescapable. A Baeyer-Villiger reaction may ensue to give the expected product and there is no evidence that such a mechanism should lead to chemilumines-cence. On the other hand the flavin peroxide is not activated by a carbonyl group and is probably incapable of accepting an electron from the assumed »activator«, the flavin nucleus. Early claims [10] that this mechanism was operating were withdrawn [11] when it was seen that no correlation of the wavelength of the emitted light could be made with the fluorescence of the product.

A related observation provides more information and virtually confirms that these compounds do not provide good models for the enzymic reaction. To obtain model peroxides in this series it is necessary to use alkyl flavinium salts to prevent the elimination of H_2O_2 which occurs in all but enzyme-bound flavin peroxides. The 5-alkyl salts provide the exact counterpart of the 4 a-peroxide believed to occur on the enzyme. However the 10 a-peroxide gives the only unambiguously identified fluorescent product. In the case of 4 a-substitution the product is not fluorescent.

Although addition of the peroxide (16) to the flavinium salt should afford the direct analogue of the natural peroxide, no significant light results from the addition.

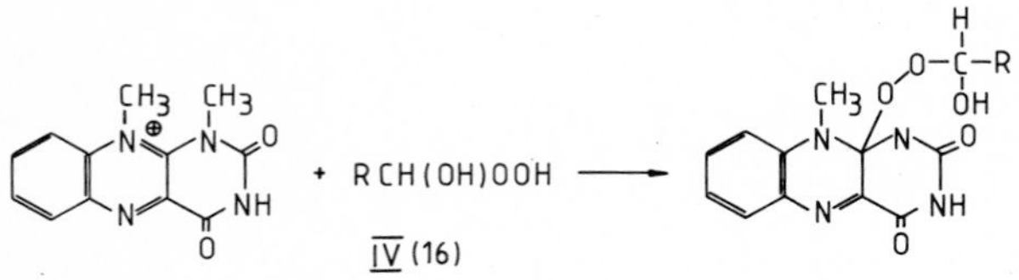

It was subsequently found that peroxides of the general structure (17) gave relatively high quantum yields.

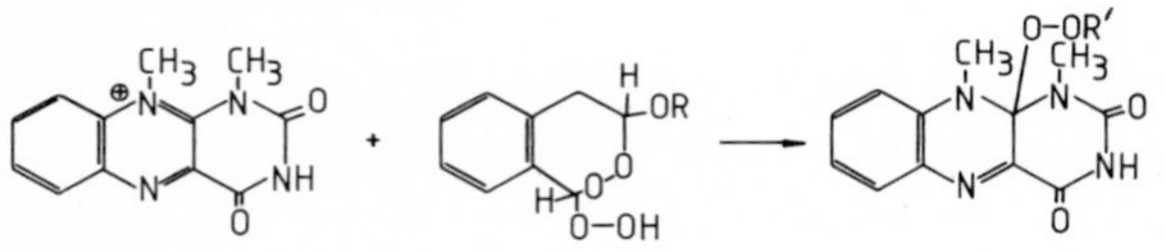

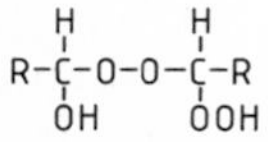

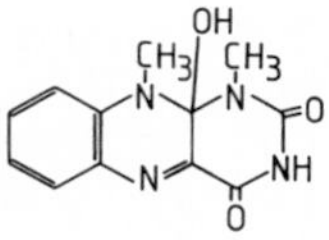

The complex peroxide can be substituted with alkyl groups and deuterium to provide a fairly clear picture of the mechanism [8]. Although there is not doubt that (18) is formed in the excited state, the original concept of activation by the electron rich flavin moiety by electron transfer to the peroxide is not valid.

This conclusion was arrived at by generalising this new chemiluminescent reaction to include a series of quinoxalinium salts (19) [12] and their adducts with peroxides of type (17).

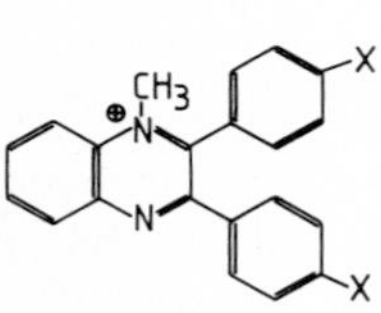

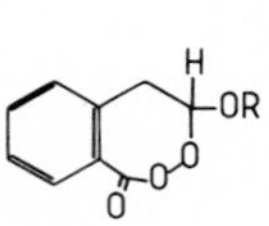

36

If electron transfer was occurring from the quinoxaline nucleus in the adduct, then a correlation between the electron withdrawing power of the groups X and the quantum yield should be observed. This was not the case. Although the light emission is absolutely dependent on having a secondary proton available, the details of the excitation step have still to be discovered.

IV.3 Dioxetanones

It might be thought that these four-membered cyclic peresters would most readily be compared with dioxetans. However there is a marked difference in their behaviour. The dioxetanones are considerably less stable than the dioxetans, and their reactions are much more susceptible to catalysis by electron donating compounds and substituents. They are key intermediates in firefly and coelenterate bioluminescence, and being a part of an electron-rich molecule, suffer decomposition as soon as they form.

Much later, reactions were devised [13] which should result in the formation of dioxetanones not substituted by electron withdrawing groups, and which were then expected to react intermolecularly with fluorescers present in the solution. The reactions of α-hydroperoxy esters (22) in alkaline solution are thought to occur via the corresponding dioxetanones.

R_1-$\overset{\displaystyle}{\underset{R_2}{}}$C—C$\overset{\displaystyle O}{\underset{OR_3}{}}$
$\overline{IV}$ (22)

R_1 = Ph–

R_2 = CH$_3$–, C$_2$H$_5$–, –CH$\overset{CH_3}{\underset{CH_3}{}}$, (aromatic ring), CH$_2$– (aromatic ring)

R_3 = CH$_3$–, C$_2$H$_5$–

Fluorescein gave the strongest emission but the quantum yield is low being only 4.2×10^{-6}. The reactions were carried out in methanol as solvent, and when water was added the rate of reaction increased about a 100-fold. Since the quantum yield decreased by the same amount, a dark reaction involving water seems to be occurring. This is in accord with an intermediate dioxetanone undergoing hydrolysis to the carboxylic acid, which would then be unable to cyclise. The initial intensity was unchanged by this addition. The relatively high efficiency of xanthene dyes such as fluorescein and eosin was ascribed to the formation of a charge transfer complex between such relatively electron rich donors and the electron deficient dioxetanone.

Synthesis of Dioxetanones
In reactions like those discussed above, the dioxetanone is only assumed as an intermediate, although isotopic studies have confirmed this assumption in the case of the luciferins. However, they have been isolated as a result of rational synthesis. The first representative was obtained by Adam [14a, 15] by singlet oxygen addition to a disilyl derivative of an ester.

IV. Chemiluminescent Peroxide Decompositions, I (exept Dioxetans)

$$(CH_3)_3C\underset{H}{\overset{}{\diagdown}}C=C\overset{-OSi(CH_3)_3}{\underset{OSi(CH_3)_3}{\diagup}} \quad \xrightarrow[(hv)]{{}^1O_2} \quad (CH_3)_3C\underset{H}{\overset{}{\diagdown}}\underset{|}{C}-\overset{O}{C}\overset{}{\diagdown}\underset{O-O-Si(CH_3)_3}{\overset{-OSi(CH_3)_3}{}}$$

$$\xrightarrow{\text{hydrolysis}} \quad (CH_3)_3C-\underset{OOH}{\overset{H}{\underset{|}{C}}}-CO_2H \quad \xrightarrow[DCC]{-H_2O} \quad (CH_3)_3C-\underset{O-O}{\overset{H}{\underset{|}{C}}}-C\overset{O}{\diagup} \qquad \text{DCC: dicyclohexylcarbodiimide}$$

An improved method [16] of preparation of the α-hydroperoxy carboxylic acid is the singlet oxygenation of α-metalated carboxylic acids.

$$\overset{R}{\underset{R'}{\diagdown}}\underset{H}{\overset{}{\underset{|}{C}}}-C\overset{O}{\underset{OH}{\diagup}} \quad \xrightarrow[THF]{LiN(C_2H_5)_2} \quad \overset{R}{\underset{R'}{\diagdown}}\underset{Li}{\overset{}{\underset{|}{C}}}-C\overset{O}{\underset{OLi}{\diagup}} \quad \xrightarrow[H^+]{{}^1O_2} \quad \overset{R}{\underset{R'}{\diagdown}}C\overset{C\overset{O}{\diagup}{}^{-OH}}{\underset{OOH}{\diagdown}}$$

Chemiluminescence ascribed to the interaction of dioxetanones was observed when singlet oxygen was passed into a solution of various ketenes in the presence of fluorescers [17]. The use of a phosphite ozonide as the source of singlet oxygen subsequently allowed the isolation of the dioxetanones themselves [18].

The availability of the compounds has allowed a comparison between them and the more studied dioxetans. Of particular interest is the likelihood that the dioxetanones in addition to forming excited carbonyl products on unimolecular decomposition, they would react intramolecularly with low ionisation potential fluorescers in a reaction related to the intermolecular case of the luciferins. Examples of both modes of reaction are discussed below.

The thermolysis of dimethyl-dioxetanone (21) revealed two different mechanisms of decomposition. In a unimolecular mechanism, the dioxetanone is cleaved into CO_2 and acetone in its excited triplet state (19):

$$\overset{CH_3}{\underset{H_3C}{\diagdown}}\overset{\diagup O}{\underset{O-O}{\boxed{}}} \quad \xrightarrow{\hspace{2cm}} \quad {}^3\ CH_3COCH_3 \ + \ CO_2$$

$$\underline{IV}\,(21)$$

The activation energy in this case varied from 87 to 93 kJ/mol in different solvents. From the temperature dependence, several competitive reaction paths for this dimethyl-dioxetanone decomposition were deduced, all having a biradical as first intermediate. Heavy-atom effects often play a role in dioxetan chemiluminescence. If DBA is used as fluorescer, the quantum yield is markedly greater than that observed when DPA is used – although the latter has a fluorescence efficiency of 0.89, compared with 0.1 for DBA. In both cases triplet-singlet energy transfer is the origin of the chemiluminescence.

Surprisingly, the yield of excited products (triplet and singlet acetone) from dimethyl-dioxetanone (21) is only 5% of that observed in tetramethyl dioxetane decomposition, although the thermolysis of (23) is ca 84 kJ/mol. more exothermic

38

than that of tetramethyl-dioxetane. SCF calculations support the mechanistic proposal depicted below [24].

$$H_3C \overset{O-O}{\underset{H_3C}{\bigsqcup}}_O \longrightarrow \left((CH_3)_2 \overset{\dot{O}}{C} - \overset{\dot{O}}{C}_{\cdot O} \right)^1 \overset{ISC}{\rightleftharpoons} \left((CH_3)_2 \overset{\dot{O}}{C} - \overset{\dot{O}}{C}_{\cdot O} \right)^3$$

$$\downarrow^{-CO_2} \qquad\qquad \downarrow^{-CO_2}$$

$${}^1(CH_3)_2 CO + {}^1(CH_3)_2 CO^{\bullet} \qquad {}^3(CH_3)_2 C = O^{\bullet}$$

$$CH_3 \overset{O-O}{\underset{CH_3}{\overset{|}{C}}} \overset{|}{\underset{CH_3}{C}} \overset{-CH_3}{\underset{CH_3}{}} \longrightarrow \left((CH_3)_2 \overset{\dot{O}}{C} - \overset{\dot{O}}{C} - (CH_3)_2 \right)^1 \overset{ISC}{\rightleftharpoons} \left((CH_3)_2 \overset{\dot{O}}{C} - \overset{\dot{O}}{C} (CH_3)_2 \right)^3$$

$$\underline{IV}(23) \qquad\qquad \downarrow^{-(CH_3)_2 CO} \qquad\qquad \downarrow^{-(CH_3)_2 CO}$$

$${}^1(CH_3)_2 CO^{\bullet} \qquad {}^1(CH_3)_2 CO \qquad {}^3(CH_3)_2 CO^{\bullet}$$

The biradicals formed from (21) and (23), respectively, are primarily produced in their singlet state, but intersystem crossing to the triplet diradical results, followed by carbon-carbon bond cleavage. As CO_2 formation from dioxetanones is more exothermic than carbonyl compound formation from tetramethyl dioxetan, it is supposed that C-C bond breakage follows more rapidly, reducing the time available for singlet to triplet conversion. As a result, the yield of excited singlet states is nearly the same in both compounds. The yield of excited *triplets*, however, is much smaller in the dioxetanone case.

An apparently bimolecular decomposition mechanism is observed in the presence of fluorescent aromatic hydrocarbons. Rubrene was the first catalyst used [25], considerably enhancing the chemiluminescence of the α-peroxy lactones. Triplet-triplet annihilation of rubrene triplets was suggested, these being produced by energy transfer from the dioxetanone decomposition products. However, this suggestion was not corroborated by further experimental evidence.

DPA, too, had a strong catalytic effect on dioxetanone chemiluminescence. The quantum yields were 20 times that of the "unimolecular" reaction and the decomposition rate considerably increased. Further investigations [21 b] had the following results.

1) The rate constant observed in the presence of fluorescers is the sum of the rate constant k_1 of the unimolecular reaction and the rate constant k_2 of the bimolecular reaction multiplied by the fluorescer ("activator") concentration:

$$k_{observed} = k_1 + k_2 \text{ [activator]}$$

2) k_2 depends on the oxidation potential of the activator (Fig. 9).
3) The chemiluminescence emission matches the fluorescence of the activator.

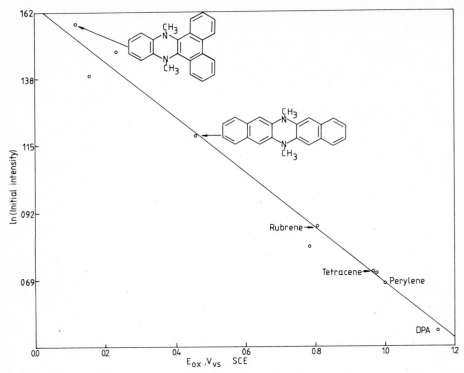

Fig. 6. Dependence of chemiluminescence intensity on oxidation potential of activator [21 b].

The experimental facts again suggest an electron-transfer mechanism.

t-Butyl-dioxetanone was found to react in an analogous way. It has been pointed out that at least in some cases the electron transfer from the activator to the dioxetan ring is *endergonic* ("up-hill"). However, stretching of the oxygen-oxygen bond of the dioxetan-activator encounter complex may considerably facilitate the formation of the radical ion pair [(dioxetan)$^{(\pm)}$/Act$^{(\pm)}$] which lies ca. 76 kJ/mol below the biradical itself [19]

IV.4 Diphenoyl Peroxide

The decomposition of this peroxide is reasonably chemiluminescent in the presence of typical fluorescent acceptors. Although the first mechanisms advanced by its discoverer [4] invoked a ground state forbidden cycloreversion, analogous to that of the dioxetans, this compound now stands as a good example of the CIEEL mechanism. It was the first in which the requirements were explicitly stated.

Thermolysis of diphenoyl peroxide in solution yields benzocoumarin and CO_2 as the main products. In the absence of easily oxidized acceptors, light yields are very low, with little evidence for the formation of triplet or singlet excited states of benzocoumarin.

Catalysis of the decomposition is observed with the rate law now characteristic of CIEEL, where D is the donor or activator:

$$k_{obs} = k_1 + k_2 [D]$$

Chemiluminescence intensity is proportional to [D], and the activation energy for light emission is derived solely from k_2. This suggests that the principal light reaction has been properly identified as the bimolecular reaction. The catalysis of the reaction can be correlated with the ionisation potential of the activator.

The reaction mechanism proposed by Koo and Schuster [4], is that of a cage charge annihilation:

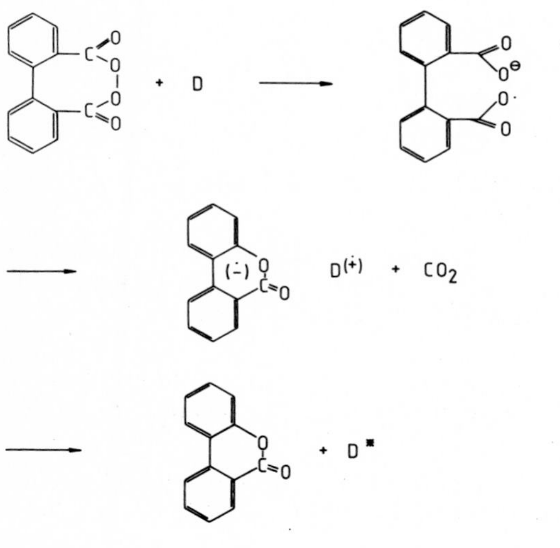

The cage nature of the reaction is confirmed by the use of diphenylamine as activator. Normally diphenylamine emits at 360 nm, but in the presence of benzocoumarin (BC), an additional peak at 450 nm, ascribed to an exciplex is seen. This 450 nm peak is only seen in chemiluminescence. As in the case of the suggestion [2] for the mechanism of active oxalate chemiluminescence, the loss of CO_2 generates a powerful reducing agent (E_o BC / $BC^{(-)}$ = −1.92 V vs SCE)

which by donating an electron forms a stable compound. Further evidence that the reaction occurs within a solvent cage comes from a study of the influence of solvent viscosity. In general the dialkyl phthalates provide the optimum viscosity and polarity for CIEEL reactions. It has been shown that with pyrene as activator only about 5% of the radical cation escapes from the cage. This is in contrast to the case of phthaloyl peroxide in which nearly 50% of the pyrene cation can be found outside the cage [22].

Direct evidence for these radical cations was provided by laser excitation of pyrene to its first excited singlet state, which then reacted with diphenoyl peroxide to give the radical cation. The cation was observed in absorption some nanoseconds after the electron transfer, time correlation being achieved by use of the exciting laser.

The second order rate constant correlates with the one electron oxidation potential of the activators.

The electron transfer is energetically favoured over the simple unimolecular formation of the peroxide diradical by 75,6 kJ/mol.

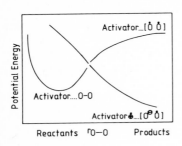

Fig. 7. Plot of potential energy vs. peroxide RO-OR bond lenght for the reaction of a peroxide with an activator (after [31]).

IV.5 Other Cyclic Diacyl Peroxides. Phthaloyl Peroxide

Other peroxides which are effective in light emission, and are superficially similar in structure to those discussed do not actually react by the same mechanism. The comparison of these two types is highly illustrative of the diversity of chemiluminescent peroxide mechanism.

Only the aromatic diacyl peroxides give rise to bright chemiluminescence. Phthaloyl peroxide generates far more light than its aliphatic, hexahydro counterpart [23, 24].

In the search for highly exergonic reactions, which might be chemiluminescent, the decomposition of cis-1,2-dihydro phthaloyl peroxide (24) was considered.

$$\Delta H \sim 504 \text{ kJ/mole}$$

IV (24)

The decomposition of the hitherto unknown cis-1,2-dihydrophthaloyl perox-ide to yield benzene and carbon dioxide is expected to release 504 kJ/mol [24 b].

It has not yet been possible to obtain (24) but during those investigations it turned out that phthaloyl peroxide (24 a) reacting with certain fluorescers, e. g. DPA, exhibited a fairly bright chemiluminescence (ca. 7% of luminol) [24 a].

Strong chemiluminescence was observed only in dimethylphthalate as sol-vent. Small quantities of water enhanced the light yield but oxygen had no distinct influence on the chemiluminescence [24 a]. Phthaloyl peroxide exhibits chemilu-minescence with a series of other fluorescent compounds such as BPEA, pyrene, perylene, 1,3-diphenyl isobenzofuran, and fluorescein.

IV (24 a)

(~ 30%)

4,5-dichlorophthaloyl peroxide, too, yielded chemiluminescence with these compounds though with a far lower quantum efficiency. Other aromatic cyclic peroxides, e. g. those based on anthracene and phenanthrene have not yet been synthesised. However, naphthalene – 2,3-dicarboxylic acid peroxide [25], (27 a) gives results with the usual activators comparable to those of phthaloyl peroxide.

Phthaloyl peroxide, in contrast to diphenoyl peroxide cannot decarboxylate readily to give a stable lactone. The corresponding lactone (25) is unreasonably strained. The as yet unknown phenanthrene dicarboxylic acid and naphthalic peroxides (26) and (27) could form stable lactone radical anions by loss of CO_2 on acceptance of an electron from the activator.

The reactions of these peroxides are clearly different from those of diphenoyl peroxide. The mechanism is likely to be complex since oxidative transformation of the added fluorescer occurs to a large extent. The main differences are that peroxides related to phthaloyl peroxide show little loss of CO_2 (about 17%) whereas the yield in the case of diphenoyl peroxide is about 80%. Quenching by oxygen is observed, but in the CIEEL pathway quenchers such as H_2O, oxygen and tetramethyl ethylene have little influence.

Both reactions are strongly catalysed by the fluorescer, but in the case of phthaloyl peroxide this is most likely to be the result of the well known induced decomposition of peroxides. This view is supported by the extensive oxidation of the fluorescer, in contrast to their truly catalytic behaviour in the CIEEL mechanism.

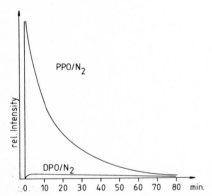

$\underline{IV}$ (24a) $\underline{IV}$(25) $\underline{IV}$(27)

$\underline{IV}$ (27a)

$\underline{IV}$ (26)

Finally, although the CIEEL mechanism is more easily understood the higher efficiency (see Fig. 8) of the more complex *phthaloyl* peroxide reactions (some 20 fold greater with comparable fluorescers and concentrations) indicates the importance of the reaction.

PPO/N$_2$

rel. Intensity

DPO/N$_2$

0   10   20   30   40   50   60   70   80   min.

Fig. 8. Chemiluminescence of Phthaloyl Peroxide (PPO) and Diphenoyl Peroxide (DPO) with 9,10- Bis(phenylethynyl)-anthracene (BPEA). Solvent: Dimethylphthalate, Temp.: 22 °C, Nitrogen Atmosphere.

The mechanism of the chemiluminescence of these peroxides has still not been fully elucidated.

The suggestion of Steinfatt [26] that dioxiranes, e. g. (28), are key intermediates in this type of chemiluminescence, is based on the chemiluminescent reaction of gem.-dihalides with hydrogen peroxide.

$\underline{IV}$ (24a) $\underline{IV}$ (28)   + fluorescer

+ oxidized fluorescer + hν

44

According to this suggestion there exists an equilibrium between the "normal" six-membered ring phthaloylperoxide (24 a) and its phthalide isomer – analogous to that of the corresponding diacyl chlorides.

$$\underline{\text{IV}}(29)$$

The isomer can be readily isolated, and actually gives rise to chemiluminescence in the presence of Na_2O_2 and an appropriate fluorescer [27]. Dioxiranes and dioxiranone (29) have been described recently [28]. It is assumed that the "isophthaloylperoxide" (28) is transformed to phthalic anhydride by transfer of an oxygen atom to the fluorescer.

Further investigations are necessary to confirm this very interesting new suggestion.

The essential differences between the two forms of peroxide reaction can be seen in the following comparison:

Diphenoyl peroxide (DPO) [19]
- non-chemiluminescent thermolysis with rate constant
  $k_1 = * \sim 8.02 \cdot 10^{-5} \, s^{-1}$
- strong increase of thermolysis rate in presence of fluorescers; chemiluminescence obsvd. = fluorescence of fluorescer
- bimolecular kinetics (1 st. order in peroxide and in fluorescer)
- fluorescer remains practically unchanged, acts as catalyst only

- CO_2 elimination ca. 80%
- »low concentration« (not defined) of H_2O, oxygen ($^3O_2$), or tetramethyl ethylene have virtually no effect on chemiluminescence

Phthaloyl peroxide [29]
- non-chemiluminescent thermolysis with rate constant
  $k_1 = * 3.8 \cdot 10^{-7} \, s^{-1}$
- strong increase of thermolysis rate in presence of fluorescers; chemiluminescence obsvd. = fluorescence of fluorescer
- bimolecular kinetics (1 st. order in peroxide and in fluorescer)
- fluorescer is transformed into many products, main product with DPA being 9,10-dihydro compound (ca. 30%, p. 43)
- CO_2 elimination maximally 17%
- in dimethyl phthalate: with PPO/BPEA saturation with $^3O_2$ yields ca. 33% of chemiluminescence obtained in N_2-saturated solution.
  With PPO and BPEA concentration = 1×10^{-3}, $\varnothing_{CL}$ is ca. $20 \times \varnothing_{CL}$ DPO/BPEA in the same concentrations

IV.6 Chemiluminescence of o-Dicarboxylic Acid Chlorides

As mentioned on p. 43 only a relatively small number of well defined cyclic diacyl peroxides has been described in the literature. One of the most frequently used methods of synthesis of those peroxides is by transformation of their bis-acid chlorides by reaction with hydrogen peroxide or an appropriate peroxide [30].

---

* Solvent DMP, under comparable conditions.

Anthracene-2,3-dicarboxylic acid can be transformed into the corresponding acid chloride with phosphorus pentachloride although chlorination in the 9 – position occurs simultaneously [31]. The question arises as to whether the peroxide (31) would be stable. Since PPO (p. 43) undergoes a catalysed chemiluminescent decomposition in the presence of fluorescers, the anthracene residue in (31) may catalyse an *intra*molecular decomposition of the diacyl peroxide group [31].

Although reaction of the acid chlorides with hydrogen peroxide and base did not lead to the isolation of definite cyclic peroxides, chemiluminescence was observed. On reacting 9,10-diphenyl anthracene-2,3-dicarboxylicacid dichloride (30) with hydrogen peroxide/tert. amine (e. g. ethyl-dicyclohexylamine-, dicyclohexylamine – or urea/H_2O_2) in dimethyl phthalate light is emitted. The emission spectrum matched the fluorescence of 9,10-diphenylanthracene-2,3-dicarboxylate (λmax. ca. 460 nm). The quantum yield is low, with ca. 10^{-5} einsteins/mol.

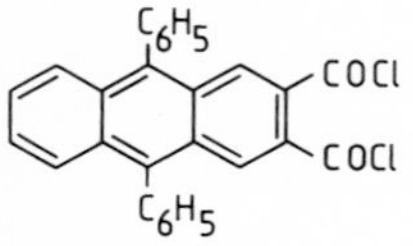

$\underline{IV}(30)$ 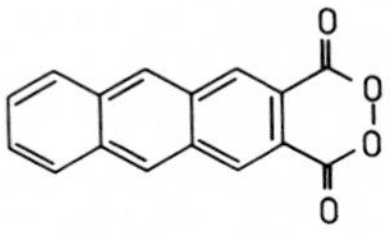 $\underline{IV}(31)$

3,6-diphenyl-3,5-cyclohexadiene-1,2-trans-dicarboxylicacid dichloride (32) was reported to yield chemiluminescence with the hydrogen peroxide complex in tetrahydrofuran [32]. As the maximum of the emission spectrum matches that of p-terphenyl fluorescence, the following mechanism was proposed:

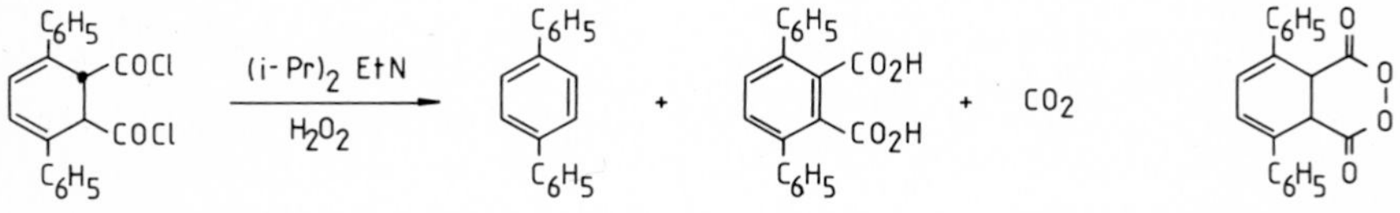

$\underline{IV}(32)$ $\underline{IV}(33)$

This is supported by energy transfer experiments with added fluorescers (perylene, rubrene, DPA) and by negative evidence for singlet oxygen involvement.

Yields of excited states were estimated by several techniques [32] to be in the range 1.5–3%. The cyclic peroxide (33) is thought to be an intermediate. Fragmentation to p-terphenyl and carbon dioxide should occur with an exothermicity of ca. 357 kJ/mol.

IV.7 Other Cyclic Peroxides

The acyl peroxides just discussed are more reactive than alkyl peroxides. Chemiluminesce only occurs at considerably higher temperatures (= ca. 150°) with the latter. Without added fluorescers the light emission is negligible.

Addition of dibenzal diperoxide (34) to a solution of dibenzanthrone (35) at 200° gives rise to a red chemiluminescence which is the emission from the first excited singlet state of the fluorescer (35) [33, 34].

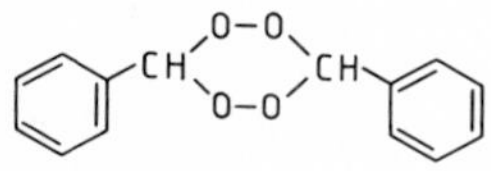

$\underline{IV}$ (34)

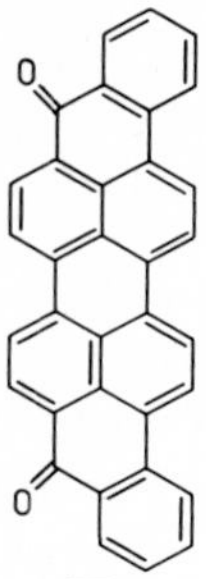

$\underline{IV}$ (35)

$\Delta H \approx 336 \, kJ \, / mol$

As dibenzal diperoxide decomposes to benzaldehyde and oxygen in a strong exothermic reaction there is sufficient energy available to excite either benzaldehyde to its triplet state (ca. 294 kJ/mol), or to produce an excited singlet oxygen species, either $O_2{}^1\Delta g$ (94.1 kJ/mol) or $^1O_2\Sigma g^+$, (156,8 kJ/mol). It was concluded, since the chemiluminescence emission came from the dibenzanthrone excited singlet state and not from the triplet, that this chemiluminescence is an energy transfer process from triplet benzaldehyde to singlet ground state dibenzanthrone.

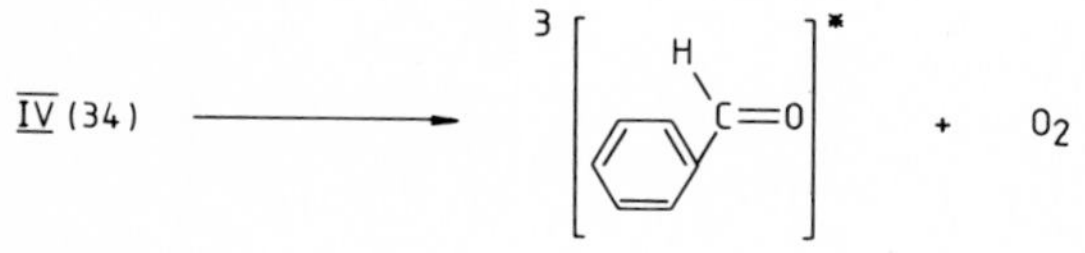

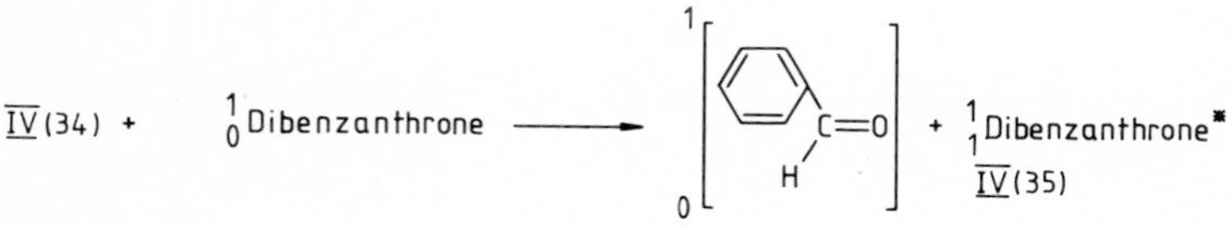

Singlet oxygen monomers, by generation of the fluorescer triplet may contribute to the excitation of dibenzanthrone. Annihilation of two dibenzanthrone triplets may lead to the excited singlet. Thus it appears to be established that the chemiluminescent reaction path does not proceed via the dibenzanthrone endoperoxide (36) originally suggested [33] as the key intermediate.

47

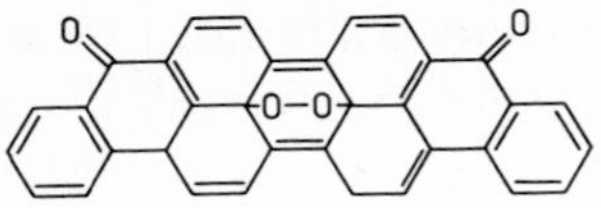

$\underline{IV}$ (36)

Instead of dibenzanthrone, other fluorescers (perylene, rhodamine B, fluorescein, DPA, anthracene) can be used. The excitation yields were found [34] to be in the range 7.74×10^{-8} to $8.76 \times 10^{-5*}$.

A linear relation was observed between the square root of the overlap integrals of the benzaldehyde phosphorescence spectrum and the absorption spectra of the fluorescers, and the number of excited singlet fluorescer molecules produced per molecule of benzaldehyde triplet. This is a strong argument for the proposed transfer mechanism.

Dibenzanthrone endoperoxide (36) may be involved in the chemiluminescent reaction of dibenzanthrone (35) with alkaline hypochlorite [34] (cf. p. 14) Chemiluminescence with quantum yields of ca 10^{-6} was also observed on heating the peroxides (37) and (38) at 150–200° in the presence of fluorescers [35].

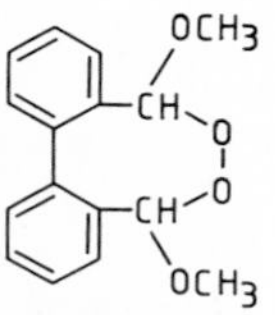

$\underline{IV}$ (37)

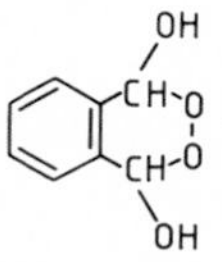

$\underline{IV}$ (38)

IV.8 Acence Endo-Peroxides

Chemiluminescence accompanying thermolysis of certain acene endoperoxides has been well known for more than 40 years. Examples are the peroxides from diphenyl anthracene (39) [36], rubrene (40) [37], and 1,4-dimethoxyl-9,10-diphenyl anthracene (42) [38, 39].

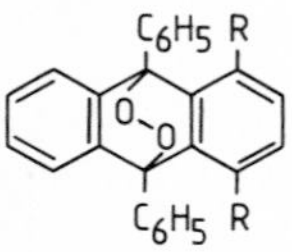

$\underline{IV}$(39)

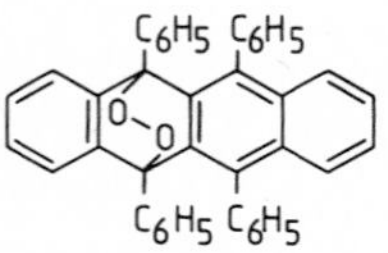

$\underline{IV}$(40)

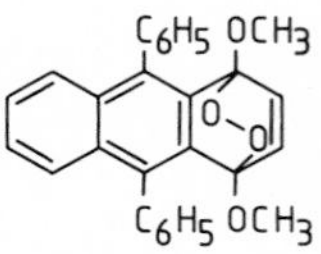

$\underline{IV}$(41)

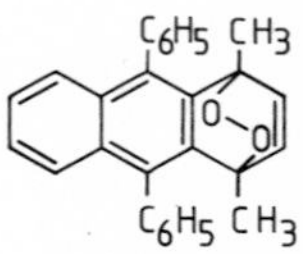

$\underline{IV}$(42)

---

* With $\emptyset_{ES} = \dfrac{\emptyset_{CL}}{\emptyset_F}$; this means a $\emptyset_{CL}$ of 1.05×10^{-5} for dibenzanthrone as fluorescer; the other fluorescers yield lower values of $\emptyset_{CL}$.

This chemiluminescence is apparently the result of energy transfer from singlet oxygen formed by decomposition of the endo-peroxide to the respective acene [40]. Indeed acene endoperoxides can be used as singlet oxygen sources [41–43]. Recent investigations on the mechanism of anthracene endo-peroxide thermolysis [44] revealed that the yield of $^1O_2$ is dependent on the structure of the endo-peroxide. The 1,4-endo-peroxides (41) and (42) produce a 95% yield of singlet oxygen, whereas the 9,10 endoperoxides (39) and (40) yield 35 and 50–75%.

Interestingly, the most efficient chemiluminescent reaction so far observed during acene endo-peroxide decomposition has been exhibited by the 1,4-endo-peroxide (41) [38]. However, this very efficient chemiluminescence evidently requires treatment of (41) with acids, and it has been shown [45] that it is not the direct decomposition of the endo-peroxide to yield (43) that produces most of the excitation energy, but the acid catalyzed cleavage of (41) to yield the compounds (44) and (45). The reaction itself had been previously described [46–47].

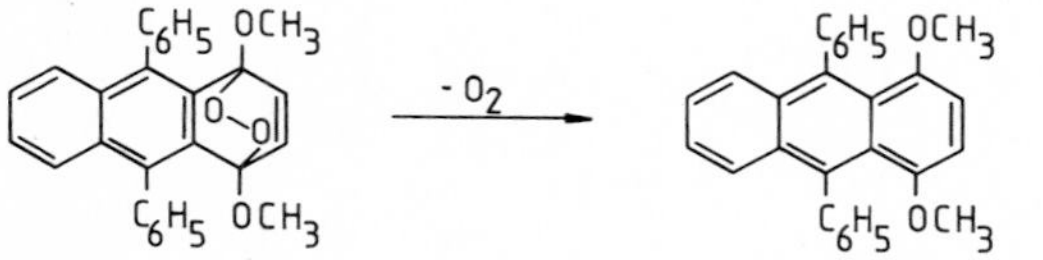

IV(41)   IV(43)

The quantum yield is in the range of 10^{-4} einsteins/mole, and is critically dependent on the acidity of the medium.

In pyridine solution there is no significant chemiluminescence. Instead (41) is decomposed to oxygen and the anthracene derivative (43) nearly quantitatively. Thermolysis of (41) in benzene, toluene or xylene also yields oxygen, together with the ester-aldehyde (44) and the long lasting chemiluminescence is reinforced by the addition of (43) and quenched by bases. The dioxetane derivative (45) is suggested as a key intermediate.

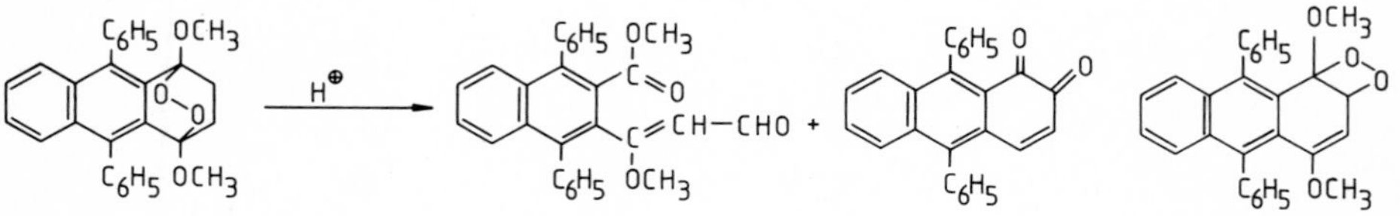

IV (41)   IV (44)   IV (45)

The emitter in the chemiluminescence of (41) is 1,4-dimethoxy-9,10-diphenyl-anthracene (43) to which the excitation energy must be transferred. It can be replaced by rubrene, but not by diphenylanthracene.

Kinetic studies on the chemiluminescence of (41) also revealed the necessity of a carboxylic acid in the reaction medium [45, 46]. The carboxylic acid cannot be replaced by hydrochloric or sulfuric acid. An unspecified intermediate, formed from the carboxylic acid and the endoperoxide may be involved. However the rearrangement to form the dioxetan will depend entirely on the strength of acid used. The dioxetan is unlikely to be stable to strong acid. Singlet oxygen may also be involved, but does not contribute greatly to the chemiluminescence.

The ester-aldehyde (44) was shown to be a reaction product of the acid treatment of (41) by high-resolution mass spectrometry [46].

IV.9 Polyhydroxy-phenol and Purpurogallin Chemiluminescence

The weak reddish chemiluminescence observed in the reaction between formaldehyde and hydrogen peroxide in an aqueous-alkaline solution of pyrogallol (Trautz-Schorygin reaction [47 a]) has been known since 1905. Singlet oxygen was suggested as the emitter, formed by recombination of peroxy radicals derived from pyrogallol [48, 49]. Recent investigations [50, 51] in this and related chemiluminescent reactions e. g. that of humic acids showed, however, that in addition to singlet oxygen chemiluminescence (Sect. II.3, p. 12) light is produced also from additional chemical reactions, of undetermined mechanism.

Pyrogallol is transformed by alkaline oxidation to a series of degradation products the most important being pupurogallin and other tropolone derivatives, which are oxidized to fluorescent products.

The chemiluminescence observed during the oxidation of humic acids is possibly related to pyrogallol or gallic acid oxidation. Humic acids are important soil constituents and the study of their chemiluminescence might be a tool in soil research. The problem here is the still very poorly defined chemical structure of humic acids.

To reduce the complexity of the problem separate investigations were performed on the oxidation by oxygen or hydrogen peroxide of polyphenols such as pyrogallol, gallic acid and aldehydes, especially formaldehyde.

Gallic acid is oxidized in aqueous-alkaline solution (pH range 8–12) to a mixture of products fluorescing in the range of 420–700 nm, the maximum being at 470–560 nm. Gallic acid itself exhibits a strong fluorescence maximum at 565 nm [52].

As the maximum of the chemiluminescence spectrum is at 640 nm, however, the main emission is ascribed to the transition

$$2 \, O_2 \, [^1\Delta_g] \rightarrow 2 \, O_2 \, {}^3\textstyle\sum g + h\nu, \; \lambda_{max} = 478 \, nm$$

as well as the excitation of the oxidation products by the oxygen collision complexes.

Structures (45–48) are suggested as intermediates.

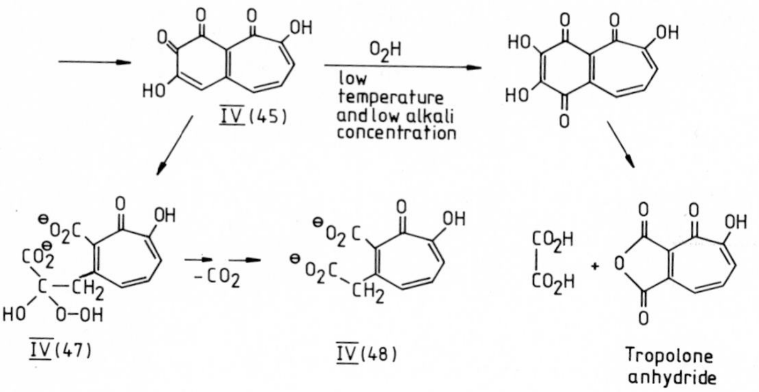

IV (45)

low temperature and low alkali concentration

IV (47)

$-CO_2$

IV (48)

Tropolone anhydride

It is a weak chemiluminescent reaction, but of interest as one of the few that emits red light.

IV.10 References

1. Marcus, R. A., J. Chem. Phys. *43*, 2654 (1965), *52*, 2803 (1970)
2. McCapra, F., Prog. Org. Chem. *8*, 258 (1973)
3. Schuster, G. B., Acc. Chem. Res. *12*, 366 (1979); Schuster, G. B., Horn, K. A. and Zupancic, J., J. Amer. Chem. Soc. *102*, 5279 (1980)
4. Koo, J-Y. and Schuster, G. B., J. Amer. Chem. Soc. *100*, 4496 (1978); *101*, 7097 (1979)
5. Linschitz, H., in Light and Life (McElroy, W. D. and Glass, G., eds.) The Johns Hopkins Press, Baltimore 1961, p. 173
6. Schmidt, S. P. and Schuster, G. B., J. Amer. Chem. Soc. *100*, 1966 (1978), *102*, 396, 7100 (1980)
7. Dixon, B. G. and Schuster, G. B., ibid. *101*, 3116 (1979); *103*, 3068 (1981)
8. McCapra, F. and Leeson, P. D., JCS Chem. Commun., 114 (1979)
9. Hastings, J. W., Methods in Enzymology (DeLuca, M. ed.) Academic Press N. Y. 1978, p. 125
10. Bruice, T. C., in Flavins and Flavoproteins (Bray, R. C., Engel, P. E. and Mayhew, S. G. (eds.) de Gruyter, N. Y. 1984, p. 45, Bruice, T. C., Acc. Chem. Res. *13*, 256 (1980)
11. Shepherd, P. T. and Bruice, T. C., J. Amer. Chem. Soc. *102*, 7774 (1980)
12. Donovan, V., Model Studies in Bacterial Bioluminescence, D. Phil. Thesis, University of Sussex 1980
13. Adam, W. and Cilento, G., Angew. Chem. Int. Ed. Engl. *22*, 529 (1983)
13. Richardson, W. H., Smith, R. S., Snyder, G., Anderson, B. and Kranz, G. L., J. Org. Chem. *37*, 3915 (1972)
14. Adam, W., Chem. i. uns. Zeit *7*, 182 (1973)

15. Adam, W. and Liu, J. C., J. Amer. Chem. Soc. *94*, 2894 (1972)
16. Adam, W., Cueto, O. and Ehrig, V., J. Org. Chem. *41*, 370 (1976)
17. Bollyky, L., J. Amer. Chem. Soc. *92*, 3230 (1970)
18. Turro, N. J. and Chow, M. F., ibid. *102*, 5058 (1980)
19. Schuster, G. B. and Schmidt S. P., Adv. Phys. Org. Chem. *18*, 187 (1982)
20. Adam, W., Simpson, G. A. and Yany, F., J. Phys. Chem. *78*, 2559 (1974)
21 a. Avramoff, M. and Sprinzak, Y., J. Amer. Chem. Soc. *85*, 1655 (1963)
21 b. Schuster, G. B. and Schmidt, S. B., in: Bioluminescence and Chemiluminescence, De Luca, M. A. and Mc Elroy, W. D. (Eds) p. 23, Academic Press, New York 1982
22. Zupancic, J. J., Horn, K. A. and Schuster, G. B., J. Amer. Chem. Soc. *102*, 5279 (1980)
23. Dervan, P. B., Jones, C. R., J. Org. Chem. *44*, 2116 (1979)
24 a. Gundermann, K.-D., Steinfatt, M. and Fiege, H., Angew. Chem. *83*, Intern. Ed. Engl. 43 (1971);
24 b. Gundermann, K.-D. and Fiege, H., 1970, unpublished results
25. Thiel, R., Dissertation, Univ. Kiel 1957
26. Steinfatt, M. F. D., J. Chem. Research (S) *1984*, 111; (M) *1984*, 1040–1062
27. Steinfatt, M. F. D., J. Chem. Research (S) *1984*, 211; (M) *1984*, 1936–1948; (S) *1985*, 140; (M) *1985*, 1673–1682
28 a. Murray, R. W. and Jeyaraman, R., J. Org. Chem. *50*, 2847 (1985)
28 b. Francisco, J. S., and Williams, I. H., Chem. Physics *95*, 373 (1985)
29. Gundermann, K.-D., Steinfatt, M., Witt, P., Paetz, C. and Pöppel, K. L., J. Chem. Res. (S) *1980*, 195; (M) *1980*, 2801
30. Kleinfeller, H., Angew. Chem. *65*, 543 (1953)
31. Steinfatt, M., Dissertation, TU Clausthal 1975
32 a. Schuster, G. B., J. Amer. Chem. Soc. *99*, 651 (1977)
32 b. Witt, P., Diplomarbeit, TU Clausthal 1974
33. Kurtz, R. B., Ann. N. Y. Acad. Sci. *16*, 399 (1954)
34. Abbott, S. R., Ness, S. and Hercules, D. M., J. Amer. Chem Soc. *92*, 1128 (1976)
35. Pöppel, K.-L., Dissertation, TU Clausthal 1982
36. Dufraisse, C. and Le Bras, J., Bull. Soc. Chim. France *4*, 349 (1937)
37. Audubert, R., Trans. Farad. Soc. *35*, 197 (1939)
38. Dufraisse, C., and Velluz, L., Bull. Soc. Chim. France *9*, 171 (1942)
39. Saito, I. and Matsuura, T., in: Wasserman, H. H. and Murray, R. W., (Eds.), Singlet Oxygen, p. 511, esp. 524, Academic Press, New York 1979
40. Bergman, W. and Mc Lean, M. J., Chem. Revs. *28*, 367 (1941)
41. Wasserman, H. H., and Keehn, P. M., J. Amer. Chem. Soc. *94*, 298 (1972)
42. Wasserman, H. H. and Larsen, D. L., J. Chem. Soc., Chem. Commun. *1972*, 253
43. Wasserman, H. H., Scheffer, J. R. and Cooper, J. L., J. Amer. Chem. Soc. *94*, 4991 (1972)
44. Turro, N. J., Chow, M. F. and Rigaudy, J., ibid. *101*, 1300 (1979)
45. Baldwin, J. E., Basson, H. H. and Krauss Jr., H., Chem. Commun. *1968*, 984
45 a. Wilson, T., Photochem. Photobiol. *10*, 441 (1969)
46. Rigaudy, J., Pure Appl. Chem. *10*, 169 (1968)
47. Rigaudy, J., Delétang, C., Partel, D. and Cuong, N. K., Compt. rend. *267*, 1714 (1968)
47 a. Trautz, M., Z. physik. Chem. *53*, 1 (1905)
47 b. Trautz, M. and Schorigin, P., Z. wiss. Phot. *3*, 80 (1905)
48. Bowen, E. J. and Lloyd, R. A., Proc. Chem. Soc. *1963*, 305
49. Gundermann, K.-D., Chemilumineszenz organischer Verbindungen, p. 115, Springer-Verlag Berlin 1968

50. Sławinska, D., Photochem. Photobiol. *28*, 453 (1978)
51. Sławinski, J., Puzyna, W. and Sławinska, D., ibid. *28*, 459 (1978) and literature cited therein
52. Sławinska, D. and Sławinski, J., Analyt. Chem. *47*, 2101 (1975)

V. Peroxide Decompositions, II: Dioxetans

Like many reactive molecules with slightly unconventional arrangements of bonds – usually involving a high degree of ring strain – dioxetans had been erroneously identified [1] long before their actual synthesis [2]. Thus there was a fair degree of resistance to their use in rationalising chemiluminescence mechanism. Although some circumstantial evidence[3] for their involvement had appeared before 1969, the application of orbital symmetry rules [4, 5] in an attempt to explain the formation of excited states provoked attempts to study them. Shortly afterwards the first synthesis [2] of an isolable dioxetan showed that although unstable, they were certainly not so reactive as to prevent manipulation. They were indeed chemiluminescent. Since then over one hundred dioxetans of varying stability have been synthesised. Compounds with half lives from minutes to thousands of years at room temperature have been obtained [6].

There are two principal synthetic routes to dioxetans. The method introduced by Kopecky and Mumford [2] is the peroxide analogue of epoxide formation.

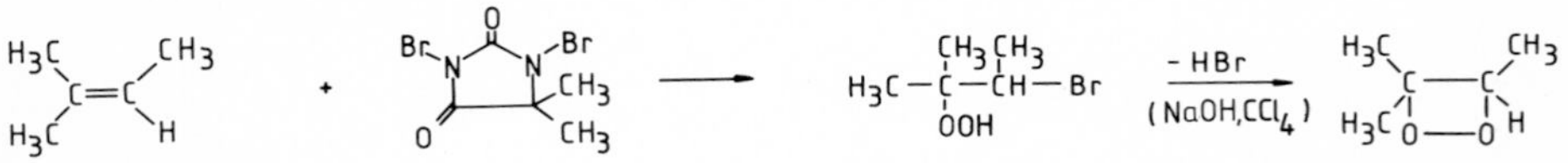

Other cyclisation reactions lead to dioxetans and dioxetanones [2, 7]. The most convenient method is reaction of olefins with singlet oxygen [8, 9, 10]. Limitations of this method include olefins with insufficient reactivity and those having allylic H-atoms capable of taking part in the "ene" reaction which leads to allylic hydroperoxides.

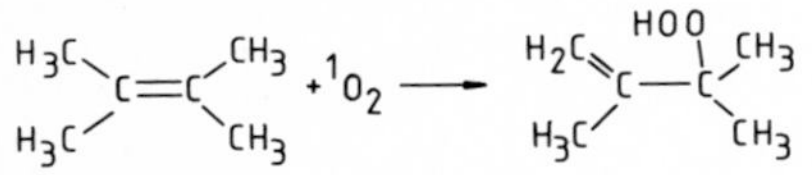

Exceptions to this rule are compounds of the diamantylidene type (59): they form 1,2-dioxetans with singlet oxygen in spite of the presence of allylic hydrogen atoms because an ene reaction would produce a hydroperoxide with a very strained double bond (forbidden by Bredt's rule).

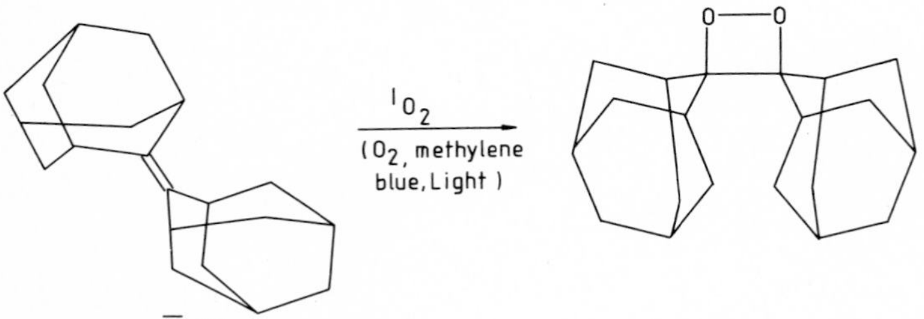

$\underline{V}$(59)

54

The singlet oxygen + olefin route is particularly effective when the double bond has electron releasing substituents.

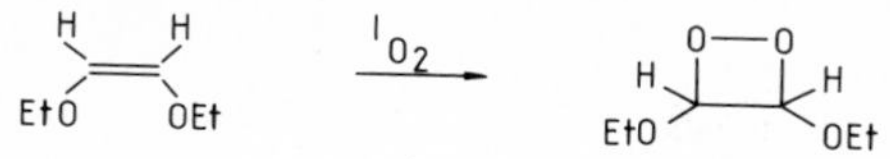

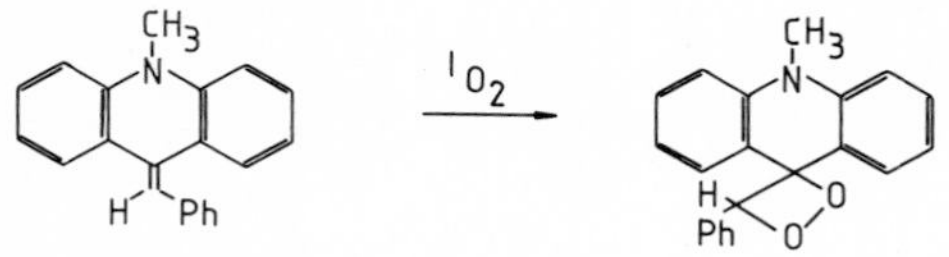

Very many dioxetans, particularly simple ones like tetramethyl-dioxetan are light yellow in colour owing to a long absorption tail which extends into the visible region of the spectrum. The properties of dioxetans have been reviewed, such reviews [2, 6, 11] being particularly useful for those interested in the physical aspects of the decomposition reaction.

As the full range of dioxetans became available, it emerged that they fell sharply into two classes. Dioxetans with alkyl aryl or other substituents ("simple dioxetans") with little electron donating power yield triplet excited carbonyl products, whereas the second type, with strongly electron donating substituents (mainly nitrogen and oxyanion) gives very high yields of singlet excitation. This latter sort is particularly significant in relation to the high singlet yields found in bioluminescence.

V.1 Simple Dioxetans

Examples of this type are shown below:

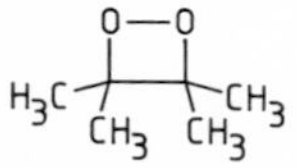

[2] 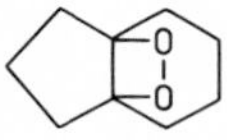   [12] 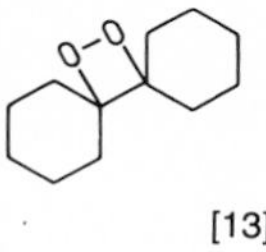   [13]

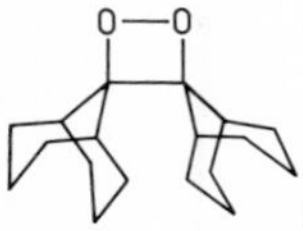

[14] 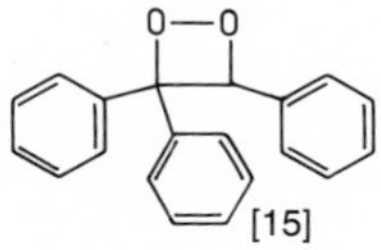   [15]

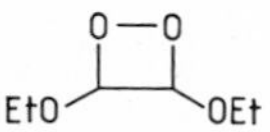

[16] 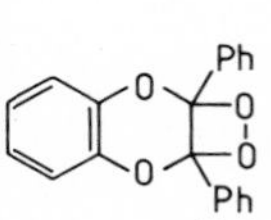   [17]

V.2 Excitation Yields and Multiplicity

The stability of these dioxetans varies from that of compounds with a half-life of minutes to indefinitely stable. Decomposition in an inert solvent, with heating if required to give an appropriate rate, is accompanied by the emission of blue light, peaking at about 400 nm. This was identified [18], in the case of tetramethyldioxetan as acetone fluorescence. Removal of oxygen with a stream of N_2 increases the light yield about 100-fold, and the maximum shifts to about 440 nm. Addition of triplet quenchers such as penta-1,3-diene severely reduces the 440 nm emission confirming that it represents acetone phosphorescence.

Phosphorescence from *photo*-excited acetone (the triplet being formed from the singlet by intersystem crossing) is very much less [19]. Thus the triplet ketone appears to be formed *directly* in the decomposition of the dioxetan. The yield of excited singlet in the thermal decomposition is almost always about 100 times less than that of the triplet.

V.3 Measurement of Singlet and Triplet Yields

Various techniques have been devised to establish the relative yield of singlet and triplet excited states, not least because such high yields of triplet excitation from a singlet ground state reactant have a strong bearing on the mechanism and will require explanation.

Russian workers [20] had developed an energy transfer technique for the measurement of excited singlet and triplet carbonyl products formed during hydrocarbon oxidation. Two fluorescent aromatic hydrocarbons, 9,10-diphenylanthracene (DPA) and 9,10-dibromoanthracene (DBA) accept, respectively, singlet and triplet energy preferentially. The heavy atom effect of the bromine atoms, by spin-orbit coupling, allows enhanced triplet to singlet energy transfer. This is a more strongly forbidden process for DPA. The resulting possibilities are

$$^1D^* + DPA \xrightarrow{k_1} D + {}^1DPA^*$$

$$^3D^* + DPA \xrightarrow{k_2} D + {}^1DPA^*$$

$$^1D^* + DBA \xrightarrow{k_3} D + {}^1DBA^*$$

$$^3D^* + DBA \xrightarrow{k_4} D + {}^1DBA^*$$

where D is the donor ketone. Only the excited singlet of the acceptor is considered since only fluorescence is detected. In principle, knowing the rate constants it should be possible to measure the absolute yields of the triplet and singlet excited ketone by this method. There is some uncertainty about the values [6], but the experiments are nevertheless very helpful. DBA is about 10 times less fluorescent at room temperature than DPA yet more light is produced from DBA. Allowance must be made for the fact that the fluorescence efficiency of DBA decreases markedly with increase in temperature. All experiments confirm the high yield of triplet carbonyl products. Europium chelates which are very efficient triplet acceptors and produce deep red light, give similar results [20, 21], although these are quantitatively less secure [6].

Since excited carbonyl triplets are excellent sensitisers of photochemical reactions, these can be used to measure excitation, particularly triplet yields.

Several of the reactions which have been used are shown below

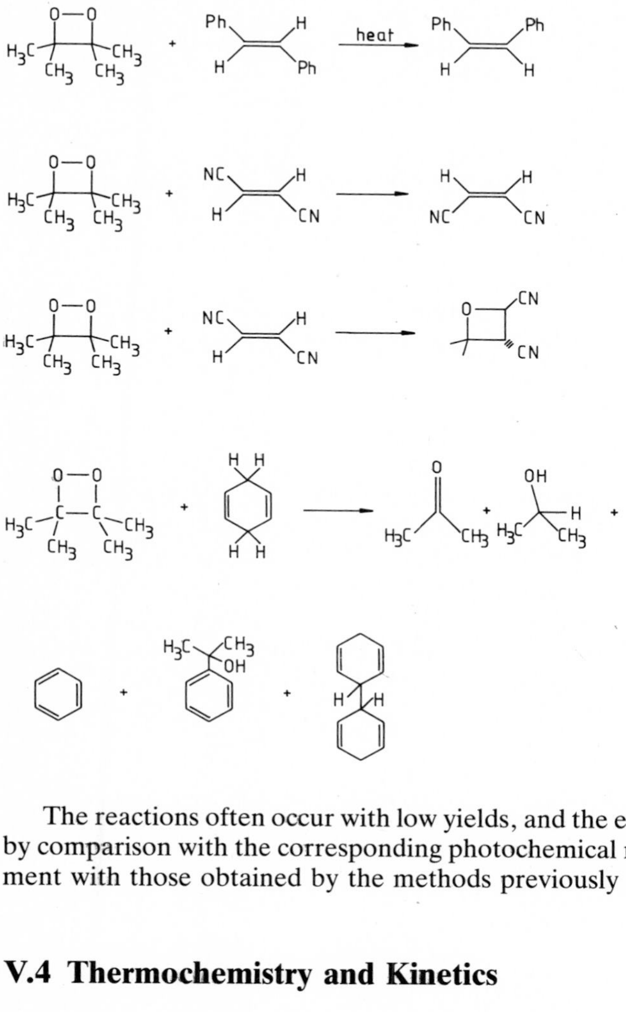

The reactions often occur with low yields, and the excitation yields calculated by comparison with the corresponding photochemical reactions give poor agreement with those obtained by the methods previously described.

V.4 Thermochemistry and Kinetics

The formation of either singlet or triplet excited state depends on there being sufficient energy, including the activation energy, to satisfy the thermodynamic requirements. Calculation of the reaction enthalpy can be made by using Benson's group additivity method, although the value for the very important strain energy can only be assumed to be equivalent to that of cyclobutane – about 109 kJ M^{-1}. The reaction enthalpy of tetramethyldioxetan has been measured [24] at

$298 \pm 12{,}6 \text{kJM}^{-1}$ for the decomposition of the solid. A lower value of $256.2 \pm 12.6\,\text{kJM}^{-1}$ in dibutyl phthalate solution is obtained. A typical activation energy is $100.8\,\text{kJM}^{-1}$ although values as high as $147\,\text{kJM}^{-1}$ have been obtained for the special case, adamantylidene-adamantane dioxetane [25].

There is apparently sufficient energy for the population of the singlet state of the carbonyl products, though only sufficient to excite one of the carbonyl fragments. The error in the calculations is hard to quantify but may be as much as 21–$42\ \text{kJM}^{-1}$. This *could* mean that there is insufficient energy to populate the singlet excited carbonyl state – a point worth bearing in mind during the subsequent discussion of the mechanism of excited state formation.

The study of the kinetic properties of dioxetan reactions is among the most convenient in the whole of physical organic chemistry. A large number of dioxetan structures have thus been synthesised for this purpose. The results have been tabulated in a very thorough review of the whole field [6]. The rate of reaction can be followed by the direct fluorescence of the products, or by energy transfer to various acceptors, particularly DBA. Arrhenius plots are easily obtained with high precision. A most useful and ingenious technique, taking advantage of the ease of following light emission was introduced by Wilson and Schaap [26]. A rapid change in temperature (usually a decrease) allows the activation energy to be obtained directly from the equation

$$\text{Ea} = \frac{\text{R}\ln\text{I}_1\text{I}_2}{\text{I}/\text{T}_1 - \text{I}/\text{T}_2}$$

where I_1 an I_2 are the initial and final intensities at the respective temperatures.

Although the activation energies obtained by the new and traditional methods are usually the same, the temperature drop method measures the light reaction *alone,* since the change in concentration during the measurement is negligible. Thus any discrepancy is highly significant, pointing to competing (dark) reactions [23, 27]. Also it is possible to measure [28] the activation energies of singlet and triplet excited state formation from tetramethyldioxetan separately by measuring both the fluorescence and phosphorescence emissions. It seems that the transition state lies above both singlet and triplet in energy, a result which is in accord with predictions based on correlation diagrams and calculations.

V.5 Mechanism of Excited State Formation from Dioxetans

The large number of investigations into dioxetans was provoked by the suggestion [3] that if the reaction were concerted, then an anti-aromatic transition state would require the formation of an excited state product. After much calculation and experiment, essentially two proposals for the excitation mechanism exist. These are the concerted and diradical intermediate mechanisms, each with its attendant explanations for the high excitation yields characteristic of dioxetans.

Although the idea of a concerted decomposition initiated work in the area, as the number of simple dioxetans increased and methods for "titrating" singlets and triplets appeared, it became obvious that triplet states predominated overwhelmingly in the products. A concerted route with no opportunity for spin inversion should of course produce only singlets as the primary product. It is still not clear, in spite of the evidence that triplets are formed *directly,* how much of the triplet yield may be the result of an enhanced rate of interruption crossing from the singlet, perhaps involving an exciplex of the two carbonyl fragments.

However other evidence particularly that produced by Richardson and his co-workers [15,30] is very much in accord with a two step, diradical process. A variety of calculations also supports this conclusion, as does evidence drawn from the stability of sterically hindered dioxetans [14]. The main features of the argument which have emerged over the last ten years are briefly summarised.

V.5.1 The Diradical Mechanism

Thermodynamic calculations [29], based on the scheme shown, with the reasonable assumptions that step 1 is rate limiting and that because of the extreme exothermicity, $k_2 \gg k_{-1}$

The results were in agreement with the trend in activation energy observed experimentally for a series of dioxetans with an increasing number of alkyl groups. The unsubstituted parent dioxetan is not sufficiently stable (as predicted) to allow its isolation, but the others gave good experimental values. This analysis does not constitute a proof of the diradical formation, but is compatible with it.

A clear distinction between the concerted and diradical routes would appear possible by placing substituents on the methylene groups so that the cleavage of the C–C bond would be enhanced by developing conjugation.

Various dioxetans of this sort, particularly with $R_1 = H$, $R_2 = R_3 = R_4 = Ph$, have been synthesised [15]. The activation energy is exactly within the range for simple alkyl substitution, and there is no effect of solvent polarity on the decomposition. The significance of this last point will be apparent after the discussion of the concerted and electron transfer mechanisms. This useful approach has been extended [30] to a study of substituent effects in the series based on (1) and (2). Although 3,3–bis (p-anisyl)-1,2-dioxetan has a lower E_a ($87.8 \pm 12.6\,kJM^{-1}$) than the

$\underline{V}$ (1) $\underline{V}$ (2) $\underline{V}$ (3)

others, this is about what would be expected for a substituent effect on the oxygen-centred diradical. Certainly the reaction constant would be very different for a rate-determining step involving C–C bond breakage. Isotope effects – substitution by deuterium – have also been used to support the diradical mechanism [31]. No isotope effect was noted for (3), the conclusion being that rehybridisation at carbon does not occur in the transition state. An *inverse* secondary isotope effect was observed [32] for the decomposition of per-deuterotetramethyldioxan. Although this does not bear directly on the diradi-cal nature of the reaction, it demonstrates that the transition state is more crowded than the reactant. This is explained by the stretching of the O–O bond.

Other substitutions confirm that the O–O bond appears to stretch before the C–C bond. This is particularly so for those dioxetans with very large rigid substituents such as adamantyl residues [14, 52]. If the C–C bond breaking appeared early in the transition state, then such crowding should lead to a lower than average stability. The reverse is very much the case.

However the view that the O–O bond simply stretches is altered by the fact [33] that 3,3-(i.e. geminal) substitution seems more important than 3,4-substitu-tion. A group of dioxetans of this sort has been synthesised with methyl and ethyl substituents so disposed as to make this point. Some workers hold the view that the transition state involves twisting about the C–C bond. Large groups by virtue of their inertia would resist this [34]. Very many bridged dioxetans have also been synthesised [35], such as (4) to (7).

$\underline{V}(4)$ $\underline{V}(5)$ $\underline{V}(6)$ $\underline{V}(7)$

The conclusion is generally in agreement with twisting in the transition state but it is not wholly convincing.

Some attempt has also been made to correlate substitution with excitation yields from these dioxetans but no clear picture emerges.

V.5.2 The Concerted or Electron Transfer Mechanism

These seemingly easily distinguished mechanisms are considered together as an alternative to the diradical mechanism for reasons which will become apparent. Electron transfer is particularly important in the more electron deficient, cyclic per-ester dioxetanones. Dioxetans in general do not respond to "activators" – fluorescent compounds of low ionisation potential (see p. 40) almost certainly because they are poorer oxidants than dioxetanones. There is however a hint that simple dioxetans, if sufficiently strained may accept electrons from an activator [36]. The light emitted from the cyclic compound (8) shows a linear dependence on DPA concentration, and the ionisation potential of several fluorescent compounds. However, true electron exchange luminescence cannot be

$\underline{V}(8)$

occurring since the *rate* of decomposition is not affected. Electron transfer cannot be occurring in the rate determining step.

The concerted mechanism for dioxetan decomposition was first proposed because the high exothermicity of the reaction and the anti-aromatic nature of the transition state could lead to crossing of ground and excited state energy surfaces. It was assumed that population of a singlet excited state of the carbonyl product would result. Since spin inversion is expected to require a much longer time to occur than is available in a concerted process, the mounting evidence that triplet products are almost exclusively formed remains the biggest objection to the acceptance of this mechanism for simple dioxetans. In addition to the convincing work already described in an attempt to establish the diradical mechanism, the decomposition of dialkyl hyponitrites demonstrates [37] fairly conclusively that cage annihilation of two alkoxyradicals can produce very high yields of triplet carbonyl products. The similarity between the two reactions has been pointed out and can be seen from the scheme:

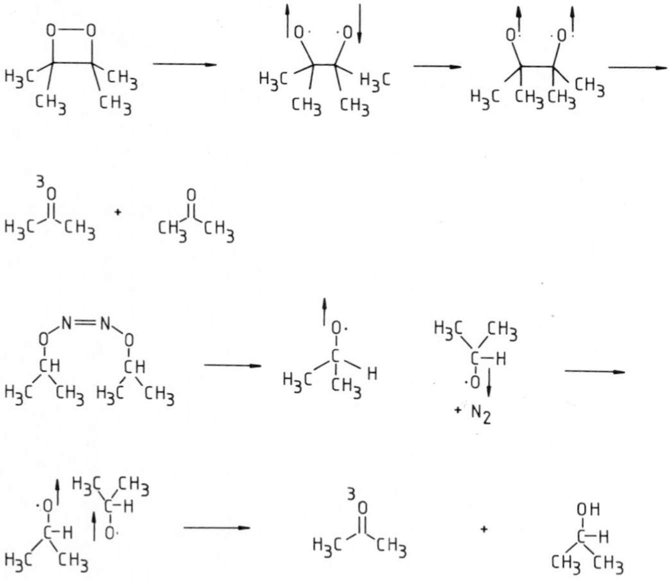

In spite of this evidence, there are calculations which show that symmetry forbidden reactions of this sort are likely to involve crossing to triplet surfaces [38]. Another suggestion is that in the cleavage of the O–O bond rotation about the C–O bond enhances the spin-orbit coupling which increases the rate of the necessary intersystem crossing from the singlet to triplet state [11].

The evidence presented for almost total O–O cleavage before the C–C bond breaks seems incontrovertible. If a concerted mechanism is to be considered, it can probably only be done in the context of a very unsymmetrical transition state. The lifetime of the diradical is certainly extremely short, and whether it has a discrete existence in all cases of simple dioxetan cleavage may never be established. Experiments in the gas phase intensify this difficulty. Absorption of infra-red radiation causes decomposition [39], but the failure to detect radicals limits their lifetime to less than 5ns. Irradiation in the ultra-violet by a pulsed laser [40], although not of such direct relevance since an electronically excited dioxetan is involved reduces this still further to 10 ps. In this latter experiment two pathways – one giving singlet excited states and the other a thermal diradical (triplet) pathway – appear to be involved. During the laser experiment an intermediate with a longer lifetime than that of acetone fluorescence was detected. If the interpretation that this represents an exciplex of the two product acetone molecules is correct, then this is a most important result. Such an exciplex would not be seen in solution since it would be readily quenched, and this may account for the low singlet yields in solution.

Although the various calculations referred to should, in principle, give an answer to the problem of radical versus concerted pathways, the several correlation diagrams published [4] provide an effective meeting point for the two mechanisms, even if less rigorous.

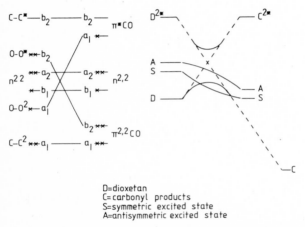

D=dioxetan
C=carbonyl products
S=symmetric excited state
A=antisymmetric excited state

Fig. 8a. State correlation diagrams for dioxetan decomposition to carbonyl compounds.

The lone pair orbitals (n) are degenerate, giving two excited state surfaces which have charge-transfer character and will be repulsive. The lower surface (S) may represent the suspected exciplex from which the excited carbonyl product escapes.

It is entirely possible that the radical and concerted paths are related. The connection may be made by noting that as the O–O bond stretches, all of the orbitals other than those of the C–C bond become degenerate not only with respect to each other but with the diradical. If substituents on the dioxetan either weaken the C–C bond sufficiently or influence the crossing on to the excited state surface, then a concerted, and most probably, a singlet state reaction will occur. Ab initio calculations show that a large number of excited states cross the reaction path, although the conclusions are not the same as the above.

Not only is there a probable connection between the radical and concerted paths, but the electron transfer pathway may also be included. It has been noted that dioxetans, unlike dioxetanones, do not show catalysed decomposition in the presence of fluorescent activators. However there is no doubt that the rate of decomposition of dioxetans with strongly electron donating substituents is unusually high. The first example of this type of dioxetan [9] was synthesised by the addition of singlet oxygen to 10,10′-dimethyl-9,9′-biacridylidene (9)

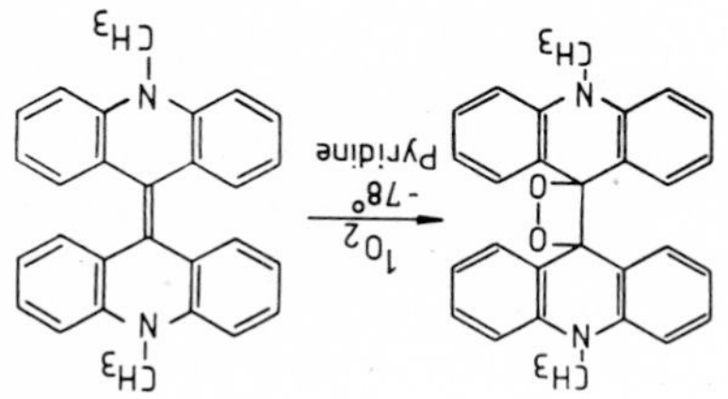

High singlet yields were obtained from its decomposition (at room temperature). Subsequent quantitative measurement gave an activation energy of $72,24\,kJM^{-1}$, much below the typical $100,8\,kJM^{-1}$ of simple dioxetans. Another related example [41] is (10), with an activation energy of $82,7\,kJM^{-1}$. Also, by taking advantage of the stabilising

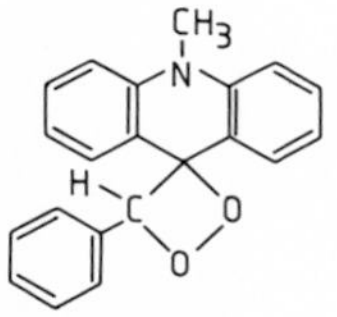

$\underline{V}$ (10)

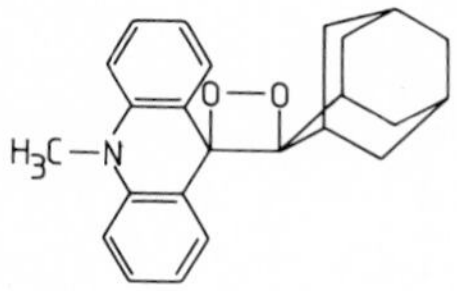

$\underline{V}$ (11)

effect of the adamantyl group [52], isolation of its relative (11) was made possible. Thus the unique character of this type of dioxetan could be established. A marked solvent effect was observed, polar solvents greatly increasing the rate of decomposition. The singlet excitation yield was a very high 12%. In addition to the solvent effect, an increase of 10^6 in rate was found when a solution of the dioxetan was treated with a solid such as SiO_2. Another dioxetan (12) had previously been shown [42] to respond dramatically to adsorption on solids, or by solution in trifluoro-ethanol. The singlet yield also increases dramatically.

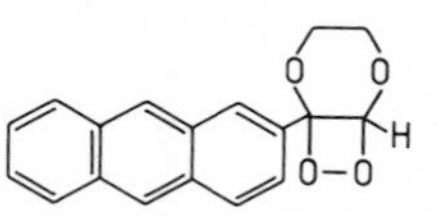

$\underline{V}$ (12)

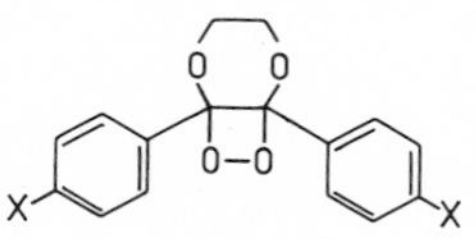

$\underline{V}$ (13)

These effects were explained by the polarisation (by adsorption to an electropositive solid site or by hydrogen bonding) of the dioxetan peroxide bond thus catalysing an electron transfer from the electron rich anthracene. It is noteworthy

63

that the naphthalene analogue [43] is a normal 'simple' dioxetan in its properties (E_a: 103,3 kJM^{-1}). In the polar solvents the sharper anthracene ester emission of the product is shifted to longer wavelength and considerably broadened. This is clearly a charge transfer spectrum and such a state may actually play a part in the increased rate of decomposition and fluorescence yield in such solvents.

Another series of compounds (13) very neatly shows the transition between the two types of behaviour [43]. Of the usual range of Hammett substituents, X, only the -NMe$_2$ grouping shows any deviation from simple behaviour. The (13) ϱ-value is small (-0.38) and the average E_a is about 101 kJM^{-1}. The dimethyl-amino-substituted compound does not lie on the line, has an E_a of 81,1 kJM^{-1}, a much higher singlet excitation yield (10^3 greater than X = H), and sensitivity to solvent polarity.

The transition between the two types can be achieved in a single compound [44] (14)

$\underline{V}$(14)

With X = H or OMe the usual triplet predominance is seen, and in its un-ionised form the compound with X = OH shows similar properties. However rapid decomposition with markedly enhanced singlet yield is seen on deprotona-tion of the phenol. This mirrors the behaviour of firefly luciferin.

Another remarkable dioxetan is (15). Decomposition in a nonpolar solvent such as n-hexane results in very short wavelength emission (320 nm) from the di-ester product [45].

$\underline{V}$(15)

When the solvent used is the more polar CH$_2$Cl$_2$, a strong second peak at 400 nm appears. Changes in the ratio of the two peaks with temperature sup-ported the proposal that the long wavelength emission originated in an exciplex of the indole ester fragment as donor and the benzoyl ester as acceptor. Fluoresc-ence at 400 nm could not be obtained by irradiation of the di-ester product, suggesting that the exciplex was uniquely formed *only* in the transition state from dioxetan decomposition. The explanation given for the different behaviour in the two solvents was that the rapid annihilation of the radical anion and cation, formed by electron transfer from the indole moiety, in a nonpolar solvent allowed insufficient time for the geometry of the exciplex to be attained. In a more polar solvent, however, the stabilisation of the radical anions gave time for the conformational change necessary for the exciplex formation.

All of the above brightly luminescent, relatively unstable dioxetans have been considered to decompose by an intramolecular electron transfer mechanism.

However there are problems associated with this extension from electrochemiluminescence and the *inter*molecular CIEEL cases.

It has already been mentioned that alkyl substituted dioxetans do not normally show exhanced rates of decomposition on intermolecular collision with low ionisation potential fluorescers. The geometric arrangement (particularly orbital overlap) of dioxetan and would-be electron donor in the compounds just discussed is less than ideal. They are in fact *simple* dioxetans with an ill-disposed electron donor. It is not clear then why they should show electron transfer from these donors (which are sometimes, e.g. in (12), considered unlikely activators in CIEEL) when they would certainly not react that way *inter*molecularly.

It is possible to describe a continuum of dioxetan reactivity by starting with the simple dioxetan with less accessible singlet states, reacting by prior O–O bond cleavage. Triplet states then result as a consequence of diradical decomposition, with a finite lifetime for the diradical that may however be close to zero. At the other end of the scale, we have dioxetans whose substitution allows the *direct* formation of a low energy *charge transfer state*. Discrete electron transfer may be an unnecessary complication in interpretation in these cases. It would thus be possible to account for the high yields of singlets, the instability of the dioxetans and the formation of exciplexes by this explanation.

V.6 Quenching Reactions of Dioxetans

Since dioxetans are the only ultimate precursors of the excitation step in chemiluminescence so far isolated, they provide a unique opportunity for the study of "dark" reactions.

Oxygen would normally be considered an effective quencher of ketone phosphorescence, yet degassing solutions of dioxetans in the presence of DBA as acceptor [26] brought an *increase* in rate and a *decrease* in quantum yield. The explanation appears to be that ^{3}DBA sensitises the decomposition of the dioxetan and ^{3}DBA survives longer in oxygen-free solution.

The longer lived, and therefore more active, triplet acetone formed in the decomposition of tetramethyldioxetan [46] causes an even more marked increase in rate by sensitised (luminescent) decomposition when oxygen is removed from the solution.

Transition metals, often found as trace impurities, catalyse the dark decomposition of dioxetans [47]. The most effective metal is $Cu^{2\oplus}$ but other metal ions ($Ni^{2\oplus}$, $Co^{2\oplus}$, $Zn^{2\oplus}$ and $Mn^{2\oplus}$) have a similar effect. The potency of the catalyst follows its Lewis acidity, suggesting a complex between the metal and dioxetan. The mechanism of the quenching of luminescence is not known.

The rare earth chelate $Eu(fod)_3$, does not catalyse the decomposition of simple dioxetans but acts as an excellent acceptor of the energy from the triplet carbonyl product. However it is an effective catalyst for the electron rich dioxetan (11), and the complex formed emits the characteristic bright red light of the europium chelate [48].

Although activated luminescence has not yet been observed with simple dioxetans, amines and olefins with low ionisation potentials are powerful catalysts for the dark decomposition of certain dioxetans [49] (e. g. (16) and (17)) but tetramethyldioxetan (perhaps because it has no electronegative substituents) is apparently unaffected by amines [6].

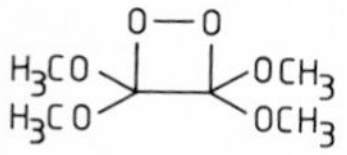

$\underline{V}(16)$

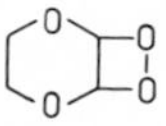

$\underline{V}(17)$

V.7 Symmetry Forbidden Reactions in Chemiluminescence

The use of orbital symmetry arguments [4] started off the interest in dioxetans, but generalisation of this application has not been extensive, nor particularly successful. Dewar benzene derivatives [50] such as (18) and (19) decompose via anti-aromatic transition states. There is only sufficient energy in the reaction of (18) to populate the triplet state of the benzene. Comparatively low yields (10^{-3} – 10^{-4}) of triplets were observed.

$\underline{V}(18)$

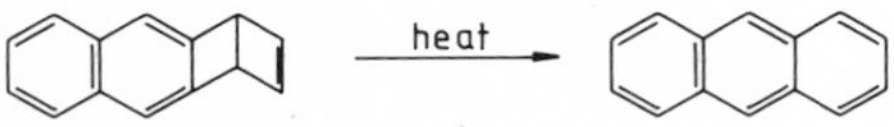

$\underline{V}$ (19)

Neither singlet nor triplet excited anthracene was observed in the second reaction shown [51]. The absence of the non-bonding oxygen orbitals and the effect on spin-orbit coupling, together with geometries unfavourable towards excited state population are all differences between hydrocarbons and dioxetans which may account for these results.

V.8 References

1a. Stephens, H. N., J. Amer. Chem. Soc. *50*, 568 (1928)

1b. Hock, H. and Schrader, H., Angew. Chem. *39*, 565 (1936)

1c. Kohler, H., Am. Chem. J. *36*, 177 (1906)

2a. Kopecky, K. R., van de Sande, J. H. and Mumford, C., Can. J. Chem. *46*, 25 (1968)

2b. Kopecky, K. R. and Mumford, C., Can. J. Chem. *47*, 709 (1969)

2c. Mumford, C., Chem. Brit. *11*, 402 (1975)

3a. McCapra, F. and Richardson, D. G., Tetrahedron Lett., 3167 (1964)

3b. McCapra, F., Richardson, D. G. and Chang, Y. C., Photochem. Photobiol. *4*, 1111

3c. White, E. H. and Harding, M. J., J. Amer. Chem. Soc. *86*, 5686 (1964)

4a. McCapra, F., Chem. Commun., 155 (1968)

4b. Kearns, D. R., Chem. Rev. *71*, 395 (1971)

5. Turro, N. J. and Devaquet, A., J. Amer. Chem. Soc., *97*, 3859 (1973)

6. Wilson, T., Int. Rev. Sci (2), *9*, 265 (1976)

7a. Adam, W. and Sakanishi, K. J. Amer. Chem. Soc. *100*, 3965 (1978)

7b. Leclerq, D., Bats, J.-P., Picard, P. L. and Moulines, J., Synthesis 778, (1982)

8. Adam, W. and Cilento, G., Chemical and Biological Generation of Excited States, Academic Press, N. Y. 1982

9a. McCapra F. and Hann, R. A., Chem. Commun., 443 (1969

9b. Lee, K., Singer, L. A. and Legg, K. D., J. Org. Chem. *41*, 2685 (1976)

10a. Bartlett, P. D. and Schaap, A. P., J. Amer. Chem. Soc. *92*, 3232, 6055 (1970);

10b. Mazur, S. and Foote, C. S., ibid. *92*, 3225 (1970)

11. Turro, N. J., Lechtken, P., Schore, N. E., Schuster, P., Steinmetzer, H.-C. and Yekta, A., Acc. Chem. Res. *4*, 97 (1974)

12. Kopecky, K. R., Lockwood, P. A., Gomez, R. R. and Ding, J.-Y., Canad. J. Chem. *59*, 851 (1981)

13. Bechara, E. J. H. and Wilson, T., J. Org. Chem. *45*, 5261 (1980)

14. Wieringa, J. H., Strating, J., Wynberg, H. and Adam, W., Tetrahedron Lett., 169 (1972)

15. Richardson, W. H., Anderegg, J. H., Price, M. E., Tappen, W. A. and O'Neal, H. E., J. Org. Chem. *43*, 2236 (1978)

16. Wilson, T. and Schaap, A. P., J. Amer. Chem. Soc. *93*, 4126 (1971)

17. Adam, W. and Cilento, G., Angew. Chem. Int. Ed. Engl. *22*, 529 (1983)

18. Turro, N. J. and Lechtken, P., J. Amer. Chem. Soc. *94*, 2886 (1972)

19. Turro, N. J. and Lechtken, P., Pure Appl. Chem. *33*, 363 (1973)

20a. Belyakov, V. A. and Vassi'ev, R. F., Photochem. Photobiol. *11*, 179 (1970)

20b. Vassi'ev, R. F., Nature (London) *196*, 668 (1962)

21. Wildes, P. D. and White, E. H., J. Amer. Chem. Soc. *93*, 6286 (1971)

22. White, E. H., Wildes, P. D., Wiecko, J., Doshan, H. and Wei, C. C., ibid. *95*, 7050 (1973)

23. Wilson, T., Landis, M. E., Baumstark, A. L. and Bartlett, P. D., ibid., *96*, 282 (1974)

24. Lechtken, P. and Höhne, G., Angew. Chem. Int. Ed. Engl. *12*, 772 (1973)

25. Lechtken, P., Chem. Ber. *109*, 2862 (1976)

26. Wilson, T. and Schaap, A. P., J. Amer. Chem. Soc. *93*, 4126 (1971)

27. Turner, N. J. and Lechtken, P., ibid., *95*, 264 (1973)

28. Steinmetzer, H. C., Yekta, A. and Turro, N. J., ibid., *96*, 282 (1974)
29 a. O'Neal, H. E. and Richardson, W. H., ibid. *92*, 6553 (1970);
29 b. O'Neal, H. E. and Richardson, W. H., ibid. *93*, 1818 (1971)
30. Richardson, W. H. and Stiggall-Estberg, D. L., ibid. *104*, 4173 (1982)
31. Koo, J.-Y. and Schuster, G. B., ibid. *99*, 5403 (1977)
32. Baumstark, A. L. and Vasquez, P. C., J. Org. Chem. *49*, 2640, (1984)
33. Baumstark, A. L. and Dunams, T., ibid., *47*, 3754 (1982)
34. Numan, H., Wieringa, J. H., Wynberg, H., Hess, J. and Vos, A., JCS. Chem. Commun., 591 (1977)
35 a. Baumstark, A. L. and Wilson, C. E., Tetrahedron Lett., 4363 (1981)
35 b. Kopecky, K. R., Lockwood, P. A., Gomez, R. R. and Ding, J-Y., Canad. J. Chem. *59*, 851 (1981)
36. Adam, W., Zinner, K., Krebs, A. and Schmalstieg, H., Tetrahedron Lett. *22*, 4567 (1981)
37. Quinga, E. M. Y. and Mendenhall, G. D., J. Amer. Chem. Soc. *105*, 6520 (1983)
38 a. Roberts, D. R., JCS Chem. Commun., 683 (1974)
38 b. Dewar, M. J. S. and Kirschner, S., J. Amer. Chem. Soc. *96*, 7578 (1974)
39. Doetschman, D. C., Fish, J. L., Lechtken, P. and Negus, D., Chem. Phys. *51*, 89 (1980)
40. Smith, K. K., Koo, J-Y., Schuster, G. B. and Kaufmann, K. J., J. Chem. Phys. *82*, 2291 (1978)
41. McCapra, F., Beheshti, I., Burford, A., Hann, R. A. and Zaklika, K. A., JCS Chem. Commun., 944 (1977)
42. Zaklika, K. A., Kissel, T., Thayer, A. L., Burns, P. A. and Schaap, A. P., Photochem. Photobiol. *30*, 35 (1979)
43. Schaap, A. P., Gagnon, S. D. and Zaklika, K. A., Tetrahedron Lett. *23*, 2943 (1982)
44. Schaap, A. P. and Gagnon, S. D., J. Amer. Chem. Soc. *104*, 3504 (1982)
45. Nakamura, H. and Goto, T., Photochem. Photobiol. *30*, 27 (1979)
46. Lechtken, P., Yekta, A. and Turro, N. J., J. Amer. Chem. Soc. *95*, 3027 (1973)
47. Bartlett, P. D., Baumstark, A. L. and Landis, M. E., ibid. *96*, 5557 (1974)
48. McCapra, F. and Watmore, D. J. Tetrahedron Lett., 5225 (1982)
49. Lee, C. S. and Wilson, T., Chemiluminescence and Bioluminescence (Cormier, M. J., Hercules, D. M. and Lee, J., eds.) Plenum, New York, 1973, p. 265
50. Lechtken, P., Breslow, R., Schmidt, A. and Turro, N. J., J. Amer. Chem. Soc. *95*, 3025 (1973)
51. Young, N. C., Carr, R. V., Li. E., McVey, K. K. and Rice, J., ibid. *96*, 2897 (1974)
52. Adam, W. and Arias, L. A., Encarnacion, Chem. Ber. *115*, 2592 (1982)

VI. Peroxy Oxalate Chemiluminescence

This type of chemiluminescence is, at present, the most efficient non-enzymatically catalysed one, [1, 2] reaching a quantum yield of 0.34 in the best case [3]. Specially formulated oxalate chemiluminescence has achieved broad practical application as an emergency light source and is commercially available (see p. 185).

This useful development began with the detection of a bluish chemiluminescence during the reaction of oxalyl chloride with hydrogen peroxide in the presence of anthracene as fluorescer [4]:

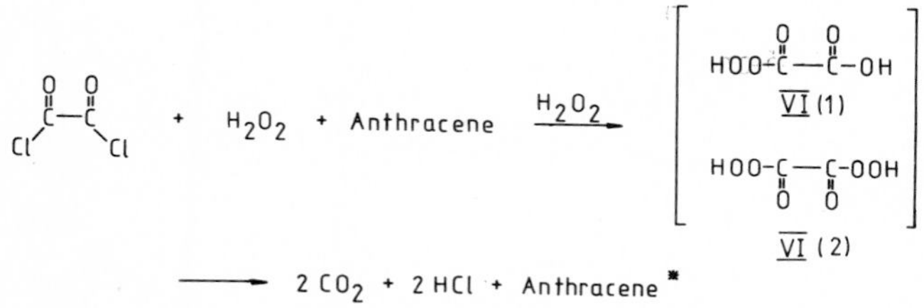

The chemiluminescence evidently was caused by some gaseous product since the fluorescence was observed when the fluorescer was placed in the vapours released from the oxalylchloride-hydrogen peroxide mixture [4]. The first interpretation, that singlet oxygen was involved was soon replaced as other oxalate derivatives were investigated.

Extensive mechanistic investigations of this reaction [5] and that of certain mixed oxalic anhydrides [6], with a variety of fluorescers resulted in new chemiluminescent systems, surpassing the best previously reported (e. g. luminol, p. 77, lucigenin, p. 109). As intermediates, at least in ether solution, mono- and di-peroxy-oxylic acid, such as (1) and (2) were proposed. The role of the added fluorescers was thought to be that of inducers of radical reactions [6].

The observation that oxalic peracids were especially efficient sources of electronic excitation energy in chemiluminescence, led to the search for other derivatives of oxalic acid capable of the formation of such peracids [2].

This resulted in the synthesis of activated oxalic esters. One of the earliest and most efficient ($\emptyset$ = 0.23 einstein/mol) is 2,4-dinitrophenyl oxalate (3):

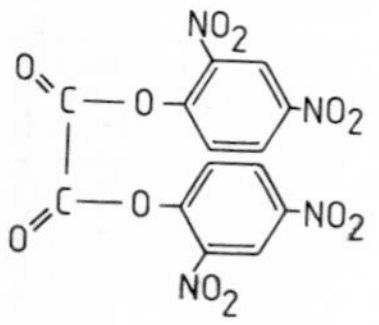

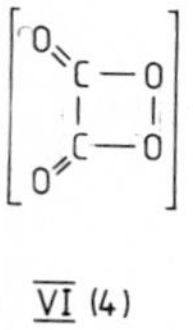

Dioxetanedione (4) was suggested as one of the possible metastable intermediates [2].

That an intermediate with a lifetime of some minutes (in the *absence* of a fluorescer) plays a key role in oxalate chemiluminescence was demonstrated by delayed fluorescer addition. Even when the fluorescer was added 70 minutes after the preparation of oxalate/hydroperoxide mixture, 53% of the quantum yield obtained by initial addition of the fluorescer was observed. On the other hand, the fluorescer evidently acts as catalyst in the chemiluminescent decomposition of that intermediate [2]. This intermediate could be transported by an inert gas stream from the oxalate/hydrogen peroxide mixture into a fluorescer solution, producing the fluorescence of the latter. In this case no consumption of the fluorescer took place. A charge-transfer complex between the fluorescer and the intermediate was proposed as early as 1967. The intermediate, as was mentioned above was assumed to be dioxetanedione (4) but it was not possible to identify it [2].

In spite of the many efforts by different groups, this situation as to the dioxetandione has not changed. The whimsical statement [7] made some years ago: "... In the reaction of oxalic esters, the mythical intermediate dioxetanedione remains as seductive and elusive as ever. The chemiluminescence community should offer a reward for its capture" is still valid at the time of writing.

VI.1 Requirements for Oxalate Chemiluminescence

The reaction has fairly strict solvent limitations. Polar aprotic solvents are required, the most usual being the dialkyl phthalates and ethylene glycol dimethyl ether. Other esters are also useful, but acetone, ethanol and halogenated hydrocarbons are much less effective. Some admixture of tert. butanol (up to about 5%) has little effect, but water and other alcohols reduce the quantum yield. It is interesting that the active site of firefly luciferase is known to be particularly hydrophobic. The mechanisms of the two reactions are rather similar in that both require a highly active ester (the adenylate in the case of the firefly). Attack by peroxide occurs in both cases (*intra*molecular in the luciferin) and this process may require a non-aqueous environment for maximum efficiency. Some of the most efficient oxalates are listed in Table 2.

In general, the aromatic residues must be substituted by strongly electron-withdrawing groups so that the resulting oxalate can easily undergo nucleophilic substitution reactions at the ester carbonyl groups.

The compound (5) belongs to a series of N-trifluoromethylsulfonyl ("triflyl") oxamides which were synthesized for similar reasons. When the nitrogen atom in.

Table 2. Electronegatively Substituted Diaryl Oxalates (from [2])

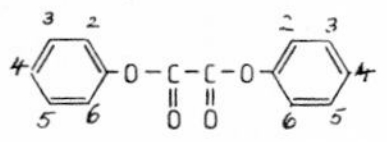

	Fluorescer
	$\emptyset_{CL}$ under optimum conditions* (einstein M^{-1})

	$\emptyset_{CL}$ under optimum conditions* (einstein M^{-1})	Fluorescer
2,4-Dinitro-	23.5	Rubrene
2,4-Dinitro-	15.4	DPA
Pentachloro-	15.1	DPA
Pentafluoro-	14.6	DPA
2-Formyl-4-nitro-	13.7	DPA
3-Trifluoromethyl-4-nitro-	12.3	DPA

Solvent: dimethylphthalate

* Concentrations: oxalate 1×10^{-2} M, H_2O_2 $0.3 - 2.0 \times 10^{-2}$ M, DPA 4×10^{-4} except in the first example where rubrene was used as fluorescer; in this case, the concentrations were: oxalate 1×10^{-3} M, H_2O_2 2.5×10^{-2} M, rubrene 5.0×10^{-4} M, benzyltrimethyl ammonium hydroxide (activator) 8.0×10^{-6} M.

the triflyl oxamides was additionally substituted by negatively substituted aryl-groups as in (5) or other compounds, some of which are listed in Table 3, very efficient chemiluminescence was produced on treatment of these compounds with hydrogen peroxide and fluorescers.

Table 3. N,N'Diaryl- N,N'-ditriflyl oxamide Chemilumi-nescence

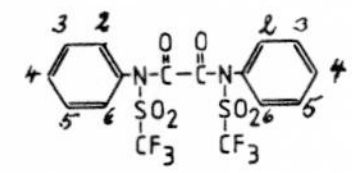

	$\emptyset_{CL}$ (einstein mol^{-1}) $\times 10^{-2}$

	$\emptyset_{CL}$ (einstein mol^{-1}) $\times 10^{-2}$
2.4.5- trichloro-	32.6–35,4
2.4- dichloro-	25.6–26.2
2-chloro-3-pyridyl-	15.5
2.4.6- trichloro-	11.4–13.6

Concentrations: oxamide 1×10^{-2}M (except the pyridyl derivative: 0.8×10^{-2}M); $H_2O_2 : 3.7 \times 10^{-1}$M; 6.75×10^{-3}M 1-chloro-9.10-bis-phenyl-ethynyl-anthracene (as fluorescer); 3×10^{-4}M sodium salicylate; solvent mixture of 75% (by volume) dibutyl phthalate, 20% dimethyl phthalate and 5% tert. butyl alcohol.

The oxalates as well as the oxamides of the types listed in tables 2 and 3 are, of course, extremely water-insoluble, and so are the fluorescers used. This is probably the main reason for the very poor efficiency of these oxalates in completely aqueous systems, although as mentioned previously water is an inherently hostile medium [8].

There are many reasons for wishing to use water as solvent. Among them are safety and analytical applications. Using the principles established in the dialkyl phthalate system a series of water soluble oxalates were synthesized, and used in conjunction with water soluble fluorescers.

Aminoalkyl substituents were introduced into the aryl oxalates as e. g., in (6):

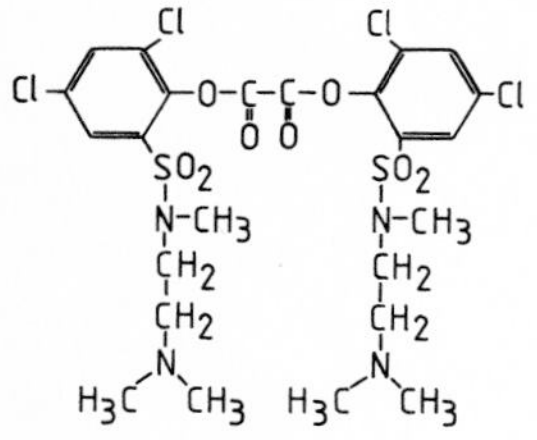

$$\overline{VI}(6)$$

The aryl groups in the triflyl oxamides were replaced by N-alkyl morpholinium-or N-alkyl pyridinium residues as in (7) and (8) [9, 10]:

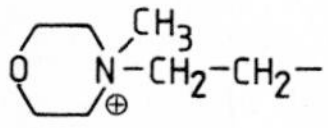

$$\overline{VI}(7) \qquad\qquad \overline{VI}(8)$$

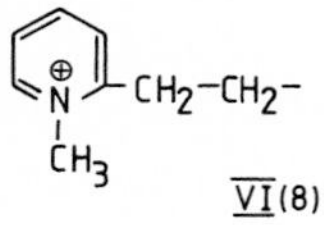

Of course the fluorescers used to transform the released energy from the reaction of hydrogen peroxide with these oxalic acid derivatives must also be water-soluble. Moreover they must be stable to attack by aqueous hydrogen peroxide. Among the fluorescers used [9, 10] are Rhodamine B, Rhodamine G perchlorate, Sulforhodamine, and 9,10-diphenylanthracene-2,6-disulfonic acid disodium salt. Table 4 lists the presently available, most efficient water-soluble

Table 4. Chemiluminescent, Water-Soluble Oxamides [10]

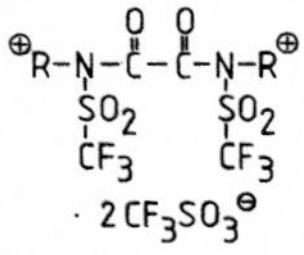	R:	Fluorescer:	percent $\emptyset_{CL}$ (einstein $M^{-1} \times 10^2$)
	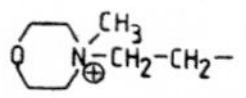	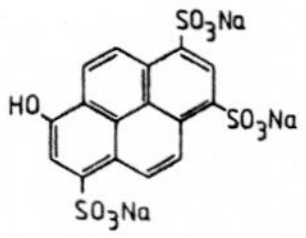	0.01

		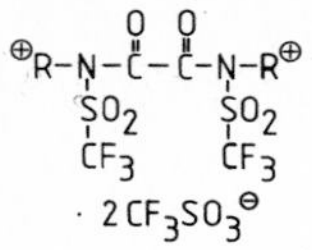 percent $\varnothing_{CL}$ (einstein $M^{-1} \times 10^2$)
R:	Fluorescer:	
	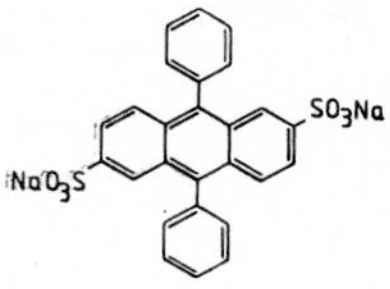 (DPADS)	0.08
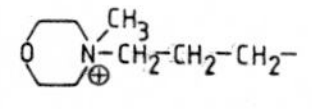	HPTS	0.01
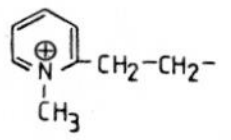	HPTS	0.002

$$\overset{\oplus}{R}-N-\overset{O}{\underset{\parallel}{C}}-\overset{O}{\underset{\parallel}{C}}-N-\overset{\oplus}{R}$$
$$\underset{SO_2}{|} \quad \underset{SO_2}{|}$$
$$\underset{CF_3}{|} \quad \underset{CF_3}{|}$$
$$\cdot \, 2\,CF_3SO_3^{\ominus}$$

Table 5. Water-soluble Oxalates [9]

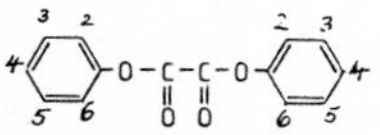

R (hydrophilizing group)	Other Substitutents
H $|\oplus$ (+) 6–SO$_2$NCH$_2$CH$_2$N (CH$_3$)$_2$ $|$ $|$ CH$_3$ H 2 Cl$^{(-)}$	2,4-dichloro-
H $|\oplus$ + 6–SO$_2$NCH$_2$CH$_2$N (CH$_3$)$_2$ $|$ $|$ C$_6$H$_{13}$ H H 2 Cl$^{(-)}$	2,4-dibromo
$|\oplus$ + 4–SO$_2$NCH$_2$CH$_2$N (C$_6$H$_{13}$)$_2$ $|$ $|$ CH$_3$ H 2 Cl$^{(-)}$	2,6-dibromo

73

oxalic acid Concentrations: Oxamide: $2-25 \times 10^{-2}$M, H_2O_2: 1.5 M, HPTS: 1.9 $\times$ 10^{-5}M-DPADS; 6.8×10^{-3}; sodium salicylate: 1,2 $\times 10^{-3}$; solvent: water. Most of the compounds described [9, 10], however, are far less efficient than the oxalic acid derivatives in aprotic media.

One has to take into account the possibility that the oxalic amide derivatives of the type (83) can form micelles, as they contain both hydrophilic and hydrophobic structures.

The analytical usefulness of water-soluble oxalates having the uniquely high efficiency of oxalates in aprotic media is evident. Use of mixed aqueous-organic solvents produces a series of complications [11], and offers no particular advantages.

VI.2 The Mechanism

The key intermediate in peroxy oxalate chemiluminescence, dioxetanedione, has not yet been isolated. Several attempts have been made to produce evidence for its existence. The gaseous products of the reaction of the oxalic ester / hydrogen peroxide reaction were fed straight into a mass spectrometer [12], and a fragment of mass 88 corresponding to C_2O_4 was identified. However, in a more recent investigation, the origin of this species was thought to be "normal" CO_2. The $C_2O_4^{(+)}$ ion is probably produced by an ion-molecule reaction involving CO_2, as the concentration of $CO_2^{(+)}$ and $C_2O_4^{(+)}$ are inversely related [13].

Another group [14] has also investigated the gaseous products from the reaction of activated oxalates with hydrogen peroxide, and compared them with those from the CO_2 afterglow [15–17], excited by microwave irradiation of CO_2. Bis- (2.4-dinitrophenyl) oxalate (3) (p. 71) and bis (pentafluoro phenyl) oxalate (9) were used for the purpose.

The gaseous products were trapped by freezing and luminesced on warming in the absence of a fluorescer. The chemiluminescence spectra derived from these experiments are shown in Fig. 9.

Reaction of the gaseous products with a lead mirror as in the Paneth technique [18] yielded lead oxide. This was interpreted as indicating the presence of $C_2O_2 \cdot$ and $CO_3 \cdot$ radicals:

$$Pb + C_2O_4 \rightarrow PbO + CO_2 + CO$$
$$Pb + CO_3 \rightarrow PbO + CO_2$$

Strong quenching effects were observed by O_2,1,3-pentadiene, tri-t-butyl phenol and tetramethylethylene (in increasing order of effectiveness). As DPA and 1,3-diphenyl isobenzofuran act as fluorescers, but not 9,10-dibromoanthracene, the active species appears to be in the singlet state, although quantitative results have not been reported. It was concluded from the kinetic and spectroscopic data that dioxetanedione as well as the 1,4-diradical are present not only in the

$$\overline{VI}(10)$$

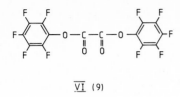

$\overline{VI}$ (9)

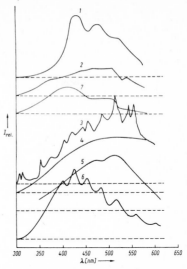

Fig. 9. Chemiluminescence spectra(corrected) of the gaseous products from the reaction of VI (9) with hydrogen peroxide, in comparison with the emission spectrum of CO_2 afterglow (after [14]).

1) Bis-pentafluorophenyl-oxalate + H_2O_2, in dimethylphthalate (DMP)
2) Spectrum of the gaseous products, expelled from the reaction mixture in 1) by Ar
3) Gaseous products, expelled from the reaction of bis (2.4-dinitrophenyl oxalate/H_2O_2 in DMP, like 2).
4) as in 3), but without Ar, at 0.1 Torr
5) as in 3), at 0.03 Torr
6) as in 3), with N_2 added subsequently
7) CO_2 afterglow, with subsequent addition of CO_2
Broken lines: spectra baselines; the intensities are not normalized.
  Spectra 2)–7) taken from the gaseous products or CO_2 without solvent. Spectrum 1) taken in DMP solution.

gaseous reaction products but also in non-aqueous solutions of the activated oxalate/hydrogen peroxide reaction. However, the radical (10) seems a very unlikely candidate indeed and would have a lifetime far too short to account for the observations.

  An interesting attempt to approach the problem from another direction has been made. Interpretation of the results is however problematical. Electrochemical oxidation [19] of oxalic acid bis (tetramethylammonium) salt in acetonitrile or DMSO using the technique of the rotating ring-disc electrode (see p. 132) and cyclic voltammetry yields carbon dioxide. The technique made it possible to infer potential intermediates in this 2-electron oxidation but no other evidence was obtained. It is assumed that the intermediates were very short lived. The following sequence was proposed:

a) $C_2O_4^{2\,(-)} - e^{\,(-)} \rightarrow C_2O_4^{(\pm)}$ (Formation of oxalate radical anion from oxalate dianion)

75

VI. Peroxy Oxalate Chemiluminescence

b) $C_2O_4^{(\pm)} \rightarrow CO_2 + CO_2^{(\pm)}$ — Decomposition of oxalate radical anion into CO_2 and CO_2 radical anion

c) $C_2O_4^{(\pm)} - e^{(-)} \rightarrow C_2O_4 \rightarrow 2\ CO_2$ — One-electron oxidation of oxalate radical anion to dioxetanedione which immediately decomposes to 2 CO_2

d) $CO_2^{(\pm)} - e^{(-)} \rightarrow CO_2$ — One-electron oxidation of CO_2 radical anion to CO_2

VI.3 References

1. For references covering the development until ca. 1968: Rauhut, M. M., Accounts Chem. Res. 2, 80 (1969)
2. Rauhut, M. M., Bollyky, L. J., Roberts, B. G., Loy, M., Whitman, R. H., Iannotta, A. V., Semsel, A. M. and Clarke, R. A., J. Amer. Chem. Soc. 89, 6515 (1967)
3. Tseng, S.-S., Mohan, A. G., Haines, L. G., Vizcarra, L. S. and Rauhut, M. M., J. Org. Chem. 44, 4113 (1979)
4. Chandross, E. A., Tetrahedron Lett. 1963, 761
5. Rauhut, M. M., Roberts, B. G. and Semsel, A. M., J. Amer. Chem. Soc. 88, 3604 (1966)
6. Bollyky, L. J., Whitman, R. H., Roberts, B. G. and Rauhut, M. M., ibid. 89, 6523 (1967)
7. Hastings, J. W. and Wilson, T., Photochem. Photobiol. 23, 461 (1976)
8. Gundermann, K.-D., Vorträge der Nordrhein-Westfälischen Akademie der Wissenschaften N 246, esp. p. 39, Westdeutscher Verlag, Köln und Opladen 1974
9. Mohan, A. G., to American Cyanamid Co, Stamford Conn., US-Pat. 4 053 430 (1977)
10. Tseng, S.-S. and Rauhut, M. M., Europ. Pat. Appl. 811 003 69.8 (1981)
11. Williams, D. C., Huff, G. F. and Seitz, W. R., Anal. Chem. 48, 1003 (1976)
12. Cordes, H. F., Richter, H. P. and Heller, C. A., J. Amer. Chem. Soc. 91, 7209 (1969)
13. De Corpo, J. J., Barovski, A., Mc Dowell, M. V. and Saalfeld, F. E., ibid. 94, 2879 (1972)
14. Stauff, J., Jaeschke, W. and Schlögl, G., Z. Phys. Chem., N. F., 99, 37 (1976)
15. Smyth, H. D., Phys. Rev. 38, 2000 (1931)
16. Gaydon, A. G., Proc. Roy. Soc. (London) 176, 505 (1940)
17. Dixon, R. N., Discuss. Farad. Soc. 35, 105 (1964)
18. Paneth, F. and Hofeditz, W., Ber. dt. chem. Ges. 62, 1335 (1929)
19. Chang, M. M., Saji, T. and Bard, A. J., J. Amer. Chem. Soc. 99, 5399 (1977)

VII. Luminol and Related Compounds

The chemiluminescence of luminol and the cyclic hydrazides of aromatic and heterocyclic acids is one of the "classical" and perhaps still most studied of chemiluminescence reactions [1–6]. The mechanism is much more complicated than that of the more recently discovered dioxetans (Chap. V). It is therefore perhaps not surprising that the latter have attracted a far greater amount of research effort in recent times since the excitation step in dioxetan chemilumines-cence offers a more acceptable interpretation of experimental results.

Water-soluble oxalates have been described (p. 72). Their efficiency is about the same as that of luminol. It will be interesting to see whether they will find the same practical, especially analytical applications. There is a long history of applications using luminol for analytical purposes (see Chap. XIII).

A better knowledge of the luminol type chemiluminescence mechanism might lead, perhaps, to a considerable increase in its chemiluminescence quantum yield, but a major attraction is surely the mystery surrounding its mechanism of reaction.

VII.1 Structure and Activity of Phthalic Acid Hydrazides

The parent of the series of which luminol is the best known member, phthalic acid hydrazide itself, provides few useful clues to the understanding of luminol chemi-luminescence. It is not chemiluminescent on oxidation by $Fe^{3(+)}$-hemin in alkaline aqueous solution. In aprotic solvents however a yellow (526 nm) emission is seen. As will be seen later, most cyclic hydrazides produce the excited state of the corresponding dicarboxylate. In this case the emission comes from the mono-anion of the starting phthalhydrazide, by back-transfer of energy from the product. One can only assume that the *primary* excited product is the dicarboxylic acid dianion, but there is not direct evidence for this [7].

Energy transfer to a linked acceptor (Sect. VII.3) helps to support this interpretation. However no information relating to the primary excitation step results.

The well-known observation concerning the effect of substituents on the phthalic hydrazide system, that electron-releasing groups increase and electron-withdrawing groups reduce the chemiluminescence quantum yield, is not yet fully understood with respect to its physico-chemical meaning. Three possibilities arise:

The substituents may be essential for the *chemical* reaction of the hydrazide oxidation itself; the excitation step may depend on them; or they may affect the fluorescence efficiency of the product. It would appear that all three influences are at work both quantitatively and qualitatively. For example, substitution may result in general instability towards oxidation as in the cases of 3,6-diamino- or 3-amino-6-methoxy phthalhydrazide [8]. The ease of oxidation on the chemiluminescent pathway is certainly significant, although it is difficult to separate this from the influence on the population of the excited state. An example of this effect is that of the 4-dialkylamino-phthalhydrazides. Differences in chemiluminescence quantum yields of up to 600-fold were observed, whereas the fluorescence quantum yields of the corresponding 4-dialkyl-amino phthalates only differ by a factor of about 10 [9].

If in the general equation for chemiluminescence quantum yield (p. 7)

$$\emptyset_{CL} = \emptyset_R \times \emptyset_{ES} \times \emptyset_F$$

$\emptyset_R$ is taken, according to the experimental evidence [10], as practically 1.0, the strongest influence of the substituent must be on the excitation step itself.

Nevertheless the fluorescence efficiency of the substituted phthalate dianion is a decisive factor in luminol type chemiluminescence. Since one can expect that the carboxylates of higher condensed aromatic hydrocarbons would exhibit high fluorescence efficiencies, a series of naphthalene-, anthracene- and homologuous o-dicarboxylic acid hydrazides have been synthesized and their chemiluminescence investigated. In the tables in Appendix, p. 205 some of these hydrazides synthesized since 1968 are listed.

As can be seen the chemiluminescence efficiency is not only dependent on the constitution of a compound, but is a property of the system as a whole. The chemical yield $\emptyset_R$ depends on the reaction conditions and the fluorescence quantum yield $\emptyset_F$ is influenced by the solvent and the presence of quenchers [9].

Therefore each hydrazide has its own optimal conditions for chemiluminescence. Among the most important parameters are hydrazide concentration and pH. With nearly all hydrazides investigated so far, there is a linear correlation between concentration and light yield only below concentrations of 1×10^{-3} M. At higher concentrations varying degrees of self-quenching are observed. This is the case with luminol. All measurements should be made in the concentration range 10^{-4} or 10^{-5} M.

Changes in pH affect several properties. One of the most important is the fluorescence efficiency of the product dicarboxylate. The most thoroughly investigated compound 3-aminophthalate (p. 82) has its maximum fluorescence yield at pH 11–11.4 (cf. p. 84). Dialkylamino-substituted phthalic or naphthalene dicarboxylic acid dianions reach their maximum fluorescence yield at a somewhat higher pH [8].

If such a dialkylamino compound is compared with luminol at pH 11, misleadingly low values will be obtained for the naphthalene derivative. The pH must also influence the ionisation of hydrogen peroxide which has a pK$_a$ of about 11.0.

The hydrogen peroxide and the hemin concentrations are not so crucial in so far as they exhibit very similar trends with all the hydrazides with no strongly

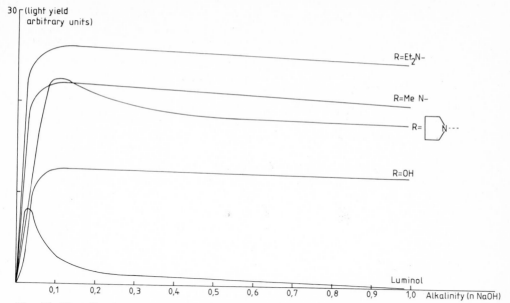

Fig. 10. Chemiluminescence efficiency of naphthalene-1, 2-dicarboxylic acid hydrazides as a function of alkalinity [12].

marked dependence on hydrazide structure [12]. If the quantum yield is measured using the luminol standard of Lee and Seliger [13], one has to decide whether to choose a "standard pH", or to find the optimal pH values, for the hydrazide under study.

Several interesting features emerge from a study of Appendix, p. 205: The most efficient compound yet synthesized is the benzoperylene hydrazide (1) $\emptyset_{CL}$ being 0.07. About 50% of the hydrazide molecules must lead to excited states on oxidation since the fluorescence quantum yield of benzoperylene dicarboxylic acid dianion is only 0.14 [14]. Interestingly the still more highly condensed coronene derivative (2) was only found to exhibit a $\emptyset_{CL}$ of 0.043 [15].

$\underline{VII}$ (1) $\underline{VII}$ (2) $\underline{VII}$ (3)

The 9,10-diphenyl anthracene derivative (3) gives only about a third of the luminol chemiluminescence efficiency, although the corresponding dicarboxylate of (3) has a fluorescence quantum yield of about 0.8, compared with a $\emptyset_F$ of 0.5 for 3-aminophthalate [16]. This result, however, was obtained for the DMSO

79

tert.-butoxide/O_2-system. When the mixed DMSO/water system (cf. Table 6 p. 84) is used, the DPA derivative is about 6 times more efficient than luminol, for which this system is not favourable. Even small quantities of DMSO added to the usual aqueous alkaline H_2O_2/hemin system produce a strong decrease in $\emptyset_{CL}$ [17].

Although the benzoperylene and the coronene derivatives (1) and (2), respectively, easily surpass luminol in chemiluminescence efficiency, suggesting an advantage of higher condensed aromatic systems, no anthracene or phenanthrene derivative has been found which even reaches the $\emptyset_{CL}$ of the best naphthalene derivatives, e.g. (4).

In substituted naphthalene dicarboxylic acid dianions, the 1,2-isomers appear to be more fluorescent than the 2,3-isomers, although the parent hydrazides (5) and (6) behave in the reverse sense [9]:

$\overline{VII}$ (4) $\overline{VII}$ (5) $\overline{VII}$ (6)

There appears to be a solvent dependence in so far as the 5-, 6- and 7-hydroxy naphthalene-1,2-hydrazides reach their highest light yields in the aprotic DMSO system whereas the dialkylamino substituted compounds are most efficient in the aqueous system. As pointed out above, it is not possible at the present time to produce even a qualitative correlation of the different substituent effects with the chemiluminescence, fluorescence, and excitation processes. Such correlations do appear to exist for other chemiluminescent compounds. For example, in the class of indolenyl peroxides [19] the efficiency of excited state formation appears to increase with decreasing excited state energy (which can be estimated from the corresponding fluorescence frequency). This also seems to be the case for luminol and its *closely related* analogues [9]. The trend in this series of substituted phthalhydrazides is similar to that found for reactants undergoing CIELL. E. H. White [20] interpreted this in terms of an electron transfer process also. Interestingly this point was made before the CIEEL hypothesis was formulated.

VII.1.1 Miscellaneous Cyclic Hydrazides

VII.1.1.1 Compounds with Multiple Hydrazide Groups

As the cyclic hydrazide groups are the essential feature in luminol type chemiluminescence, the introduction of more than one of these structures in an appropriate aromatic system may increase the chemiluminescence quantum yield.

However only one of the compounds of this type has as yet reached the efficiency of luminol, although the first representative of this series, pyromellitic hydrazide (7), described as early as 1937 [21], exhibits a "rather strong" yellow chemiluminescence. The maximum emission was found to be at 540 nm in a recent investigation, $\emptyset_{CL}$ being ca. 0.0002 einstein/mol in the aqueous system. This corresponds to about 1–2% of the efficiency of luminol. However, introduction of an amino group in (7) as in (7 a) produces

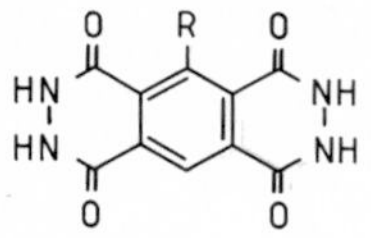

VII(7)   R = H

VII(7a) R = NH$_2$

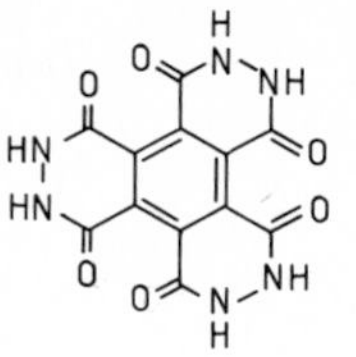

VII(8)

an enormous increase in $\emptyset_{CL}$: (7a) chemiluminesces about 3 times as strongly as luminol [22].

Mellitic acid trihydrazide (8) gives about 2,3% of the luminol chemiluminescence. It was synthesized (as its tris-hydrazinium salt) recently [23] λ_{max} 4,75, 530 nm.

"Diluminyl" (9) is a biphenyl derivative with non-coplanar benzene rings, due to steric hindrance. The chemiluminescence quantum yield is about 30% of that of luminol [18, 24, 25].

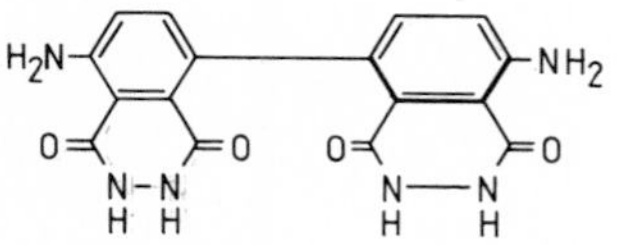

VII(9)

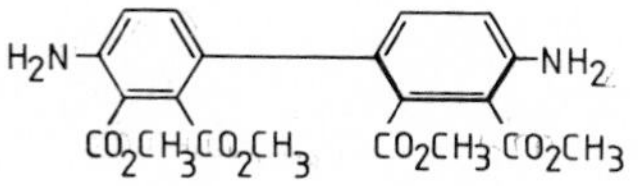

VII(10)

The fluorescence quantum yield of the corresponding tetramethylester (10) is ca. 80% of that of the double amount of the methyl 3-amino phthalate fluorescence. The relatively low chemiluminescence efficiency probably does not arise from a sort of "internal concentration quenching" but perhaps from an incomplete reaction of (9) with the oxidant. The compound will behave as a specially substituted phthalhydrazide, only one of the hydrazide groupings reacting at any given moment.

A similar discrepancy between the fluorescence efficiency of the aminophthalic ester derivative and the chemiluminescence of its corresponding hydrazide was observed in the case of oligomers of 3-amino-6-vinyl phthalic esters and the hydrazide prepared from it.

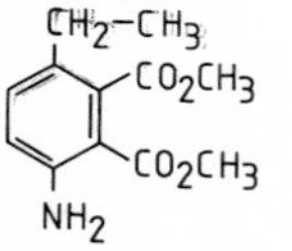

VII(11)

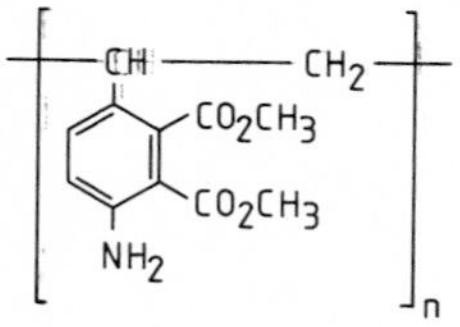

VII(12)

The polymeric ester (12) exhibits a fluorescence quantum yield of ca. 80% of that of 3-amino-6-ethyl-dimethylphthalate (11) which can be regarded as the "monomer unit" of (12). Significant self-quenching is evidently not occurring [26].

On treatment of (12) with hydrazine, a product was formed which contained ca. 70% cyclic hydrazide and 30% N-amino phthalimide groups so that the "statistical" formula is

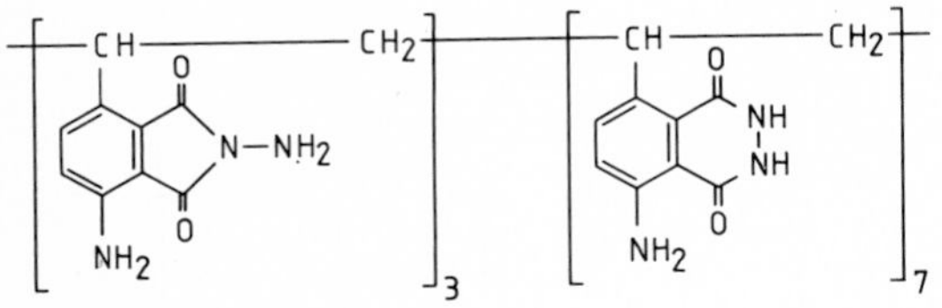

$\overline{VII}$ (13)

The chemiluminescence light yield of (13) was found to be only 0.01% of that of the "monomer", in the aqueous as well as in the DMSO system [26]. As is seen from the fluorescence efficiency of the ester (12) concentration quenching due to the "proximity" of the hydrazide groups does not seem to be the cause of the poor chemiluminescence. Nonetheless, in the compounds (14a–14c), in which the luminol groups are separated from each other by a longer hydrocarbon chain, the chemiluminescence efficiency reaches 70–82% of that of luminol [26, 27].

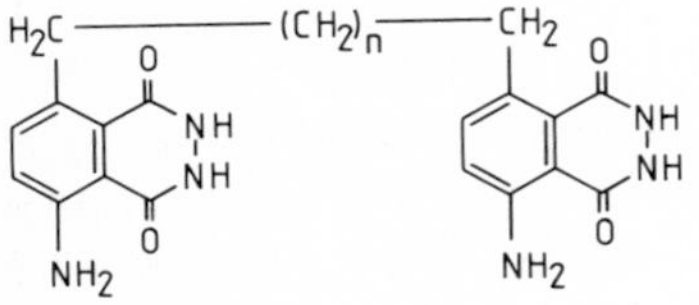

VII (14a), n = 5
VII (14b), n = 7
VII (14c), n = 9

$\overline{VII}$ (14)

Further studies of this phenomenon are evidently necessary. Various other cyclic hydrazides have been synthesized to explore other phenomena such as energy transfer and solvent effects. These, with analytical applications are treated in the relevant later sections.

VII.2 The Emitting Species

In all chemiluminescence reactions of cyclic diacyl hydrazides investigated so far the respective (substituted) phthalate dianions are the emitters.

The maximum of luminol chemiluminescence in aqueous medium is at 425 nm and in aprotic media (DMF, DMSO etc.) at 480 nm. In DMSO/water mixtures both maxima are observed [5]. These different maxima are ascribed to different species of the aminophthalate dianion (15) and (16).

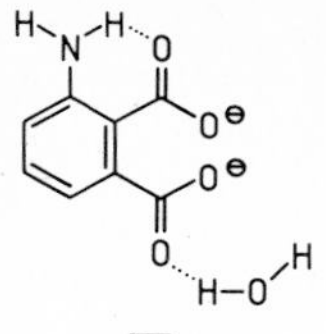

VII (15) 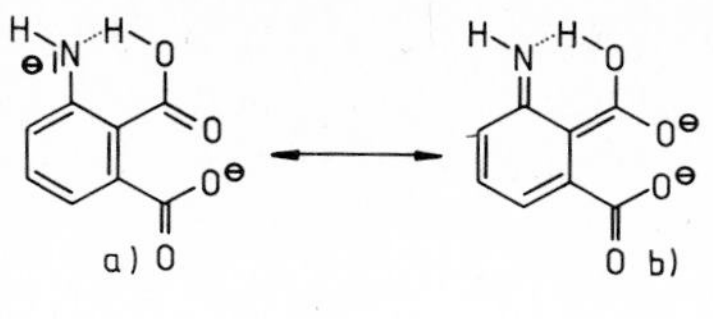   VII (16)

There is thought to be a contribution from the quinonoid form (16b), since such species are known to fluoresce at longer wavelength (about 510 nm) [5].

A strong influence is exerted by the base used in the chemiluminescent oxidation. It was observed [38–40] that only when quaternary ammonium hydroxides are used as base in the luminol reaction does there exist a one-to-one correlation between luminol chemiluminescence and aminophthalate fluorescence. With alkali metal hydroxides (NaOH, KOH) in DMSO, containing 10% water, the 425 nm maximum is more pronounced in 3-aminophthalate fluorescence than in luminol chemiluminescence.

This is explained [28] by ion pair effects. The metal ions form ion pairs with the dicarboxylate form (15) inhibiting proton transfer to give (16) and depressing the contribution of (16b). Therefore, the higher the alkali metal ion concentration, the less pronounced is the long wavelength band. This rather complicated situation can be made still more difficult, since sometimes the commercially available or freshly prepared luminol samples may contain impurities rapidly formed by air oxidation. These impurities can give spurious wavelength shifts.

On the basis of chromatographic investigations, the results mentioned above were challenged, and it was suggested that the 425 and the 480 nm luminol emission were caused by impurities in the luminol preparations, and not by different species of luminol itself [31].

This statement has not been corroborated. It demonstrates, however, the absolute necessity of luminol purification – not only in investigations concerning reaction and excitation mechanism, but also when luminol is used as a standard for quantum yield measurements [30].

3-Aminophthalic acid, on standing in aqueous solution, is in part transformed into the phthalimide derivative (17).

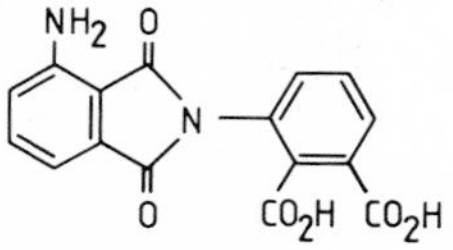

VII (17)

This product was also isolated from commercially available 3-aminophthalic acid as a yellowish impurity. Luminol, on slow oxidation in the presence of air, yields 3-aminophthalimide [32].

The mechanism is not known:

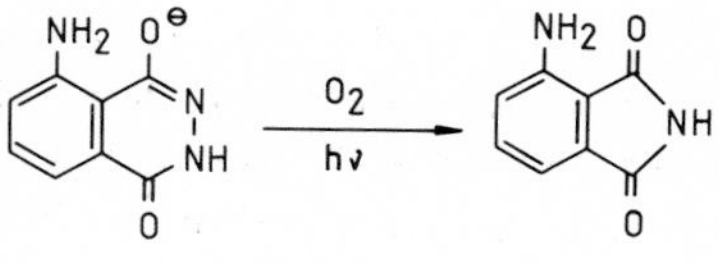

As 3-amino phthalic acid (APA) is a crucial compound in luminol chemiluminescence, very careful investigations were performed to determine its fluorescence as a function of solvent and of pH, as well as the correlation between APA fluorescence and luminol chemiluminescence [33]. See Table 6:

Table 6. Chemiluminescence maxima of luminol and fluorescence maxima of 3-amino-phthalate dianion (after Lee and Seliger [33])

Solvent	Chemiluminescence of luminol		Fluorescence of 3-amino phthalate	
	λ_{max} (nm)	ν'	λ_{max} (nm)	ν'
Water	431	23250	431	23250
DMSO	502	19900	495	20200
DMF	499	20050	497	20100
Acetonitrile (AN)	500	20000	500	20000
THF	496	20150	500	20000
DMF:AN (7:3 vol%, $-50\,°C$)	472	21200	478	20900

The quantum yield of 3-amino phthalic acid fluorescence is extremely pH-dependent in the range of pH 7–11 (in aqueous solution). $\emptyset_F$ is about 0.3, decreasing strongly with falling pH values to 0.01 in the acid region. The decrease of 3-amino phthalate fluorescence at pH values beyond 11 is probably caused by the formation of a tri-anion (18)

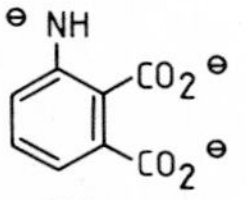

VII (18)

owing to further deprotonation. The tri-anion (18) appears to be very weakly fluorescent. Measurements of APA UV-absorption spectra lead to the conclusion that in the pH range below 11, different ionic species occur with very different fluorescence properties.

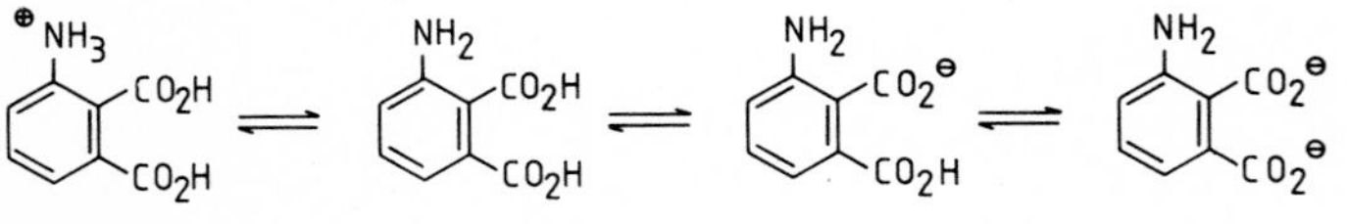

The fact that (sterically non-hindered) dialkylamino -benzene- or -naphthalene dicarboxylic acid hydrazides do not exhibit a strong decrease of chemiluminescence quantum yield beyond pH 11, is a consequence of the fact that compounds like (4) – (p. 80) – cannot form a tri-anion [34, 35].

In summary, there is no reasonable doubt that 3-aminophthalate dianion and the dicarboxylates corresponding to other hydrazides of the luminol type are the emitters in luminol type chemiluminescence.

Some discrepancies between the fluorescence spectrum of 3-aminophthalic acid and the luminol chemiluminescence spectrum are at least in part due to reabsorption of the shorter-wavelength emission by luminol monoanion [30].

VII.3 Energy Transfer in Luminol Type Chemiluminescence

The chemiluminescence of luminol is a "direct" chemiluminescence, i.e. the reaction product aminophthalate is the primary excited species. A series of compounds has been synthesized since 1967 in which the energy-generating hydrazide group and the light emitting group (a fluorescent residue) are separated [5, 36–38].

Typical examples are the N-methylacridone derivative (19), the carbazole- and benzocarbazole derivatives (20) and (21), in which the fluorescent, heterocyclic part of the molecule is linked with the phthalhydrazide structure via a bridging methylene group.

A mesomeric interaction between the donor and the fluorescing (acceptor) part of the molecule is thus not possible.

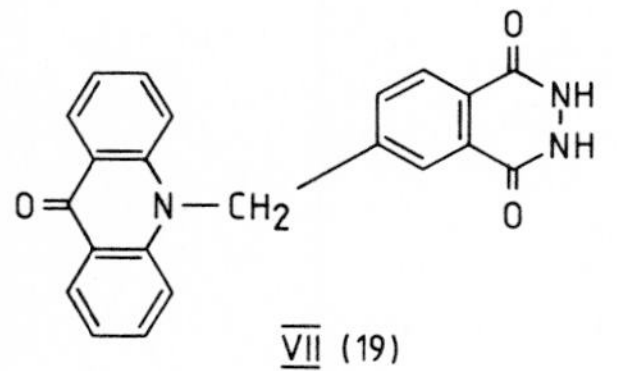

$\underline{VII}$ (19) 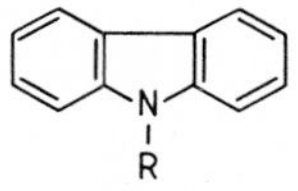 $\underline{VII}$ (20) 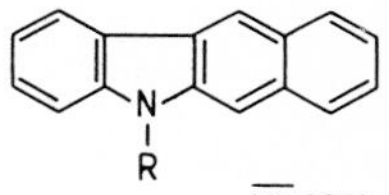 $\underline{VII}$ (21)

R:

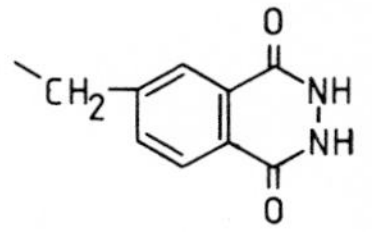

The 9,10-diphenylanthracene derivative of naphthalene-2,3-dicarboxylic acid (22), has also been examined.

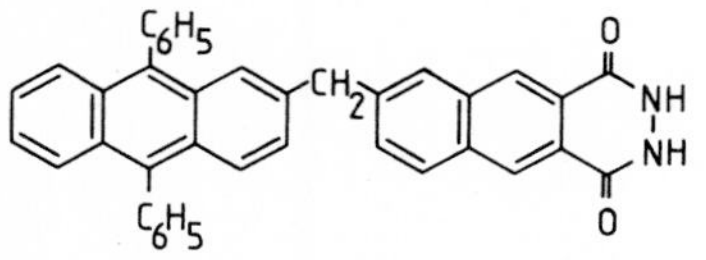

$\underline{VII}$ (22)

That the donor and the fluorescing part of the molecules in (19) and (22) actually are separated can be demonstrated. As mentioned above unsubstituted phthalic acid hydrazide does not chemiluminesce, neither does 4-methyl phthalic hydrazide [36]. A mixture of phthalic acid hydrazide and N-methyl acridone gives no light on oxidation. The conjugate (19) chemiluminesces in the aprotic system with 8% of the luminol quantum yield (ca. 1×10^{-3}), the emission matching

N-methyl acridone fluorescence. Similarly, naphthalene-2,3-dicarboxylic acid hydrazide has a very low chemiluminescence efficiency (about 5×10^{-4}), and a mixture of this hydrazide with DPA (both 10^{-4} M) yields no more light than does the hydrazide alone.

Again, linking these two groups via a methylene group as in (22) led to a compound which on oxidation chemiluminesced with an efficiency of 26% of that of luminol (ca. $3,3 \times 10^{-3}$). The chemiluminescence spectrum of (22), taken from the aqueous system, exhibits the bands of DPA fluorescence and also of naphthalene-2,3-dicarboxylate. The latter is not represented in the chemiluminescence spectrum in the aprotic system, probably due to the fact that naphthalene-2,3-dicarboxylate fluorescence in DMSO is very weak [38].

From these and additional results from other compounds of this type it was concluded that the chemiluminescence of (19) and (22) is caused by intramolecular triplet-singlet energy transfer of a mixed dipole-dipole and exchange character. This is demonstrated by the fact that the fluorescence from the singlet-state of the "emitter-moiety" is not governing the chemiluminescence of these compounds and that, moreover, the most important excited state of the phthalate moieties is the first excited triplet state [38]. Whereas in the compounds (19) and (22) the hydrazide group and the "fluorescer" group can alter their relative spatial position by (possibly restricted) rotation around the CH_2-bonds, this is not possible in paracyclophane-type compounds like (23).

H_2C ─⟨◯⟩─ CH_2
H_2C ─⟨◯⟩─ CH_2
O =⟨ ⟩= O
N–N
H H
VII (23)

Thus it was of interest to explore the behaviour of such compounds since the dipoles of both donor and acceptor transitions are unambiguously oriented with respect to each other.

If one divides the cyclophane (23) as in the preceding compounds (19) and (22), into "donor" and "fluorescer" parts it is evident that in (23) there is no fluorescer part, p-xylene being non-fluoresent. 3,6-Dimethyl phthalic hydrazide, on the other hand, is only extremely weakly chemiluminescent ($\emptyset_{CL} = 10^{-6}$).

Nevertheless, (23) chemiluminesces with an efficiency of ca. 2% of that of luminol in the aqueous – alkaline H_2O_2 – hemin system. The emission maximum was found to be 390 nm which corresponds to the fluorescence of paracyclophane-2,3-dicarboxylate [39, 40]. Thus in this the simplest member of the series a composite fluorescer is produced. The same observations can be made concerning the corresponding naphthalene analogue (24).

The would-be acceptor, 1,4-dimethylnaphthalene, is not fluorescent in the visible range of the spectrum. However, (24) again chemiluminesces on oxidation in the aqueous system, with an efficiency of 1% of that of luminol (ca. 1×10^{-4}).

The anthracenophane (25) has an efficiency of about 20% of that of luminol in the aqueous system. In DMSO/*tert*-butoxide/O_2 it exhibits the same light yield as luminol under the same conditions.

In (25) there is a true fluorescer moiety in the cyclophane system, and it represents a donor-acceptor complex system, whereas (24) and (23) very probably are forming exciplexes on oxidation [41]. The higher efficiency of this "paracyclophane energy transfer" in comparison with the methylene-linked "energizer" and "fluorescer" as in (22) is seen from the fact that in (22) the DPA-residue, having a fluorescence quantum yield of nearly unity exhibits a chemiluminescence efficiency of 26% of that of luminol whereas in (25) with the 1,4-dimethyl anthracene fluorescer ($\emptyset_{Fl}$ ca. 0.30) a light yield of 100% luminol [41] is obtained.

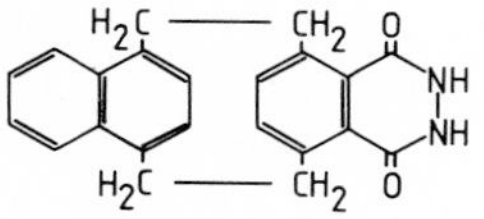

VII (24) 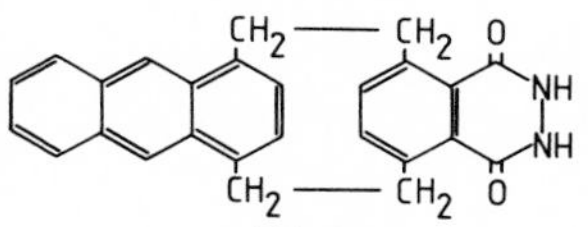   VII (25)

The chemiluminescence of maleic hydrazides substituted – via a vinyl group or directly – with the highly fluorescent DPA residue as in (26 a) could be explained by means of a CIEEL mechanism as indeed could the results described above.

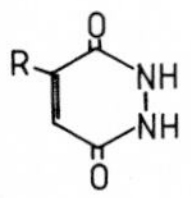

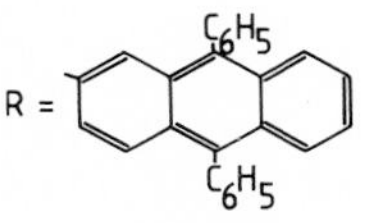

VII (26a)

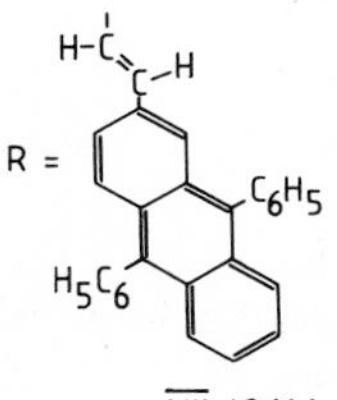

VII (26b)

The light yields were 10% (26 a) and 8% (26 b) of that of luminol* in a mixed DMSO/40% water solvent with alkali/H_2O_2/hemin. The emission maxima of the chemiluminescence spectra were 484 and 457 nm respectively (DPA has λ max. 425 nm) [42]

For comparison 9,10-diphenylanthracene-2,3-dicarboxylic acid hydrazide (3)

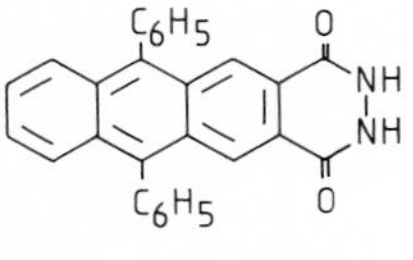

VII (3)

has a chemiluminescence efficiency of ca. 30% relative to luminol [43] (Appendix p. 207).

---

* (In the DMSO/*tert*-butoxide/O_2-system).

VII.4 Solvent Effects

The chemiluminescence quantum efficiency of luminol is practically the same in anhydrous DMSO/base/O_2 and in the aqueous alkaline H_2O_2/hemin system [33]. Addition of small amounts of water to the DMSO system drastically diminishes the chemiluminescence light intensity. In a mixed solvent of DMSO/28 mole % water/potassium hydroxide/oxygen the relative intensity was only 1/30,000 of that in the DMSO/t-butoxide/O_2 system [43 a].

The same is true if small quantities of DMSO are added to the above mentioned aqueous oxidation system [17]. These two effects are not necessarily related, since the aqueous system requires a catalyst.

There are, however, certain cyclic diacyl hydrazides which chemiluminesce very poorly in anhydrous DMSO as well as in water, but which, in mixed DMSO/water solvents, yield a rather strong chemiluminescence, e.g. 3-(2-dialkyl-aminovinyl) phthalhydrazides (see Appendix, p. 205) [44] and trans-4-'dialkyl-amino stilbene-2,3-dicarboxylic acid hydrazides [18].

Benzoperylene hydrazide (1), the most strongly chemiluminescing hydrazide,

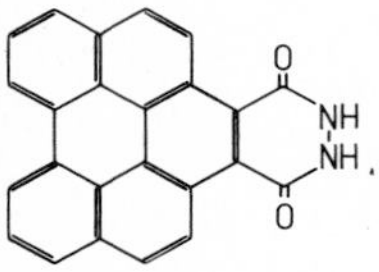

VII (1)

see Appendix p. 207) also requires a DMSO/water solvent [14] for maximum light yield. This "DMSO-effect" is not yet understood.

VII.5 Effect of Micelle-forming Agents

Addition of the micelle-forming cetyltrimethylammoniumbromide (CTAB) to solutions of 3-monoalkylamino phthalhydrazides (27) – (30) in aqueous alkaline hydrogen peroxide/potassium persulfate [42 b, 45]

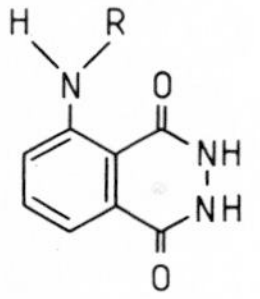

(27) R = CH_3
(28) R = C_3H_7
(29) R = C_4H_9
(30) R = C_8H_{17}

VII (27-30)

caused a moderate increase in the chemiluminescence efficiency. The maximum increase of 75% was obtained with (29) (R = n-butyl). On the role of micelles as models for luciferase see Chap. XII, p. 155.

VII.6 Catalyst Activity

As was mentioned catalysts are necessary for luminol chemiluminescence in aqueous media. When a rapid stream of carbon dioxide is passed through a basic solution of luminol containing hydrogen peroxide and manganous chloride/sodium chloride as catalyst, the intensity of the emitted chemiluminescence light passes through 4 maxima before the reaction stops (having then reached a pH of 8–9 [31].

This phenomenon very probably is caused by the successive formation of four catalytically active forms of manganese complexes differing in their coordination spheres [47]. When hemin is used instead of manganese under otherwise the same reaction conditions, there is only a single maximum in the intensity.

(see also Chap. XIII: inorganic analysis, p. 169)

VII.7 The Mechanism of Luminol Chemiluminescence

The starting point for luminol oxidation is clearly an anion, since base is an essential catalyst. Luminol monoanion (37) is present in aqueous alkaline solutions, almost as the sole ionic luminol species. Luminol dianion (38) on the other hand, has been shown to exist in aprotic solutions (DMSO or DMF, e. g.) with tert.butoxide as base [6, 29, 48].

It is not present to any significant extent in aqueous solutions.

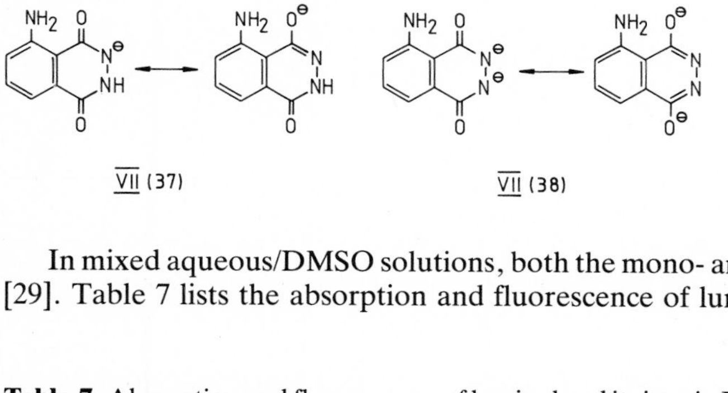

VII (37)   VII (38)

In mixed aqueous/DMSO solutions, both the mono- and dianion are observed [29]. Table 7 lists the absorption and fluorescence of luminol and its ions [29].

Table 7. Absorption and fluorescence of luminol and its ions in DMSO (from Gorsuch and Hercules [29]).

	Absorption $\lambda_{max\ nm}$, (ε)			Fluorescence $\lambda_{max\ nm}$
Luminol (neutral)	297	(7560);	360 (7720)	410
Luminol-Monoanion[1]	333	(6370);	370 (6320)	500
Luminol-Dianion[2]	310	(3250);	393 (6500)	520

[1] Sodium salt with a trace of potassium-*tert*-butoxide.
[2] *tert*-butoxide used as base.

It is therefore possible that the first steps of the chemiluminescence mechanism are different in these two classes of solvents.

In the aprotic systems, molecular oxygen can attack the dianion to form a peroxide and additional catalysis is not required for chemiluminescence.

This is almost certainly the result of the greater stability of the intermediate radical anion as opposed to the radical

VII (39)   VII (39 a)

which would be formed from the protonated mono-anion on electron abstraction by triplet O_2.

VII.7.1 Ionic Species Derived from Luminol

In protic systems – mostly water – oxygen alone is not sufficient but catalysts such as $Co^{2(+)}$, $Cu^{2(+)}$, $Cr^{3(+)}$ and many other transition metal ions, often as complex compounds such as $K_3Fe(CN)_6$ or hemin, must be present.

Irradiation with different types of high energy rays, pulse radiolysis or acoustical cavitation also cause luminol chemiluminescence in aqueous solution. All these physical sources are known to produce HO. radicals by homolysis of the water molecules. The most reasonable assumption is that in aqueous solution the necessary luminol species is a radical anion (39) which rapidly reacts with an appropriate oxidizing species either as the neutral species or after ionization.

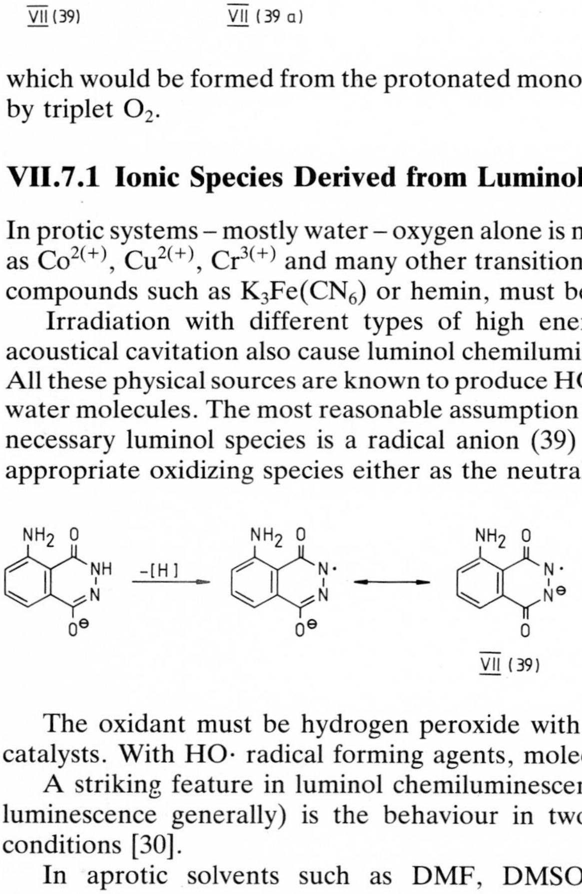

VII (39)

The oxidant must be hydrogen peroxide with metal ion or metal complex catalysts. With HO· radical forming agents, molecular oxygen is sufficient.

A striking feature in luminol chemiluminescence (and indeed in hydrazide luminescence generally) is the behaviour in two markedly different sets of conditions [30].

In aprotic solvents such as DMF, DMSO, or tetrahydrofuran, high chemiluminescence quantum yields are observed. The excitation yield $Ø_{ES}$ is about 0.09. This efficiency is not influenced to any considerable extent by temperature, polarity of solvent, or by quenchers. No catalyst other than very strong base is required. The oxidant is molecular oxygen.

In aqueous solution a variety of oxidative catalysts are required in addition to hydrogen peroxide. The results vary remarkably with choice of catalyst. Greatest efficiency is obtained with hemin as catalyst, in the pH range 11–13 ($Ø_{CL} = 0.012$–

0.013, $\emptyset_{ES} = 0.04$). These values are fairly independent of temperature, solvent viscosity, quenchers and luminol concentration.

In aqueous systems using other oxidants, such as sodium hypochlorite, potassium ferricyanide or potassium persulfate the chemiluminescence quantum yields are reduced to 0.003–0.007, probably the result of concurrent "dark" reactions. A distinct influence of temperature and luminol concentration is observed. It should be noted that the concentration range concerned is 1×10^{-3} M and below, i.e. the luminol concentration range where self-quenching does not occur [48 a].

Table 8. Luminol chemiluminescence quantum yields in different solvents with some oxidative systems (after Lee and Seliger [30])

Luminol concentration (M)	Solvent	T °C	pH	Reaction conditions	Chemiluminescence quantum yields
10^{-3}	Water	3	11,6	H_2O_2/Hemin	0,0123
10^{-3}	Water	20	11,6	H_2O_2/Hemin	0,0124
10^{-3}	Water	40	11,6	H_2O_2/Hemin	0,0135
10^{-3}	Water	54	11,6	H_2O_2/Hemin	0,0116
10^{-3}	DMSO	25	–	O_2/t-BuOK	0,0124
10^{-3}	DMSO	45	–	O_2/t-BuOK	0,0123
10^{-5}	Water	20	12,2	$K_3Fe(CN)_6$	0,00010±0,0000
10^{-5}	Water	20	11,6	NaOCl	0,004±0,001
10^{-3}	Water	20	11,6	$K_2S_2O_8$	0,007±0,001
10^{-5}	Water	20	11,6	Photosensitized with methylene	0,003
10^{-6}	Water	50	11,6	blue / O_2	0,01

Singlet oxygen is included as an oxidant, too, in Table 8. Interestingly the chemiluminescence efficiency reaches that of the H_2O_2/hemin system when the luminol concentration is low (10^{-6}).

VII.7.2 Diazaquinone Chemiluminescence

Although diazaquinones could be considered in a separate class of chemiluminescent compounds their importance lies in their association with the mechanism of luminol chemiluminescence. It is not yet certain whether they are intermediates under all reaction conditions, but they are clearly implicated.

In the first publications on luminol chemiluminescence [49] the diazaquinone "dehydroluminol" (40) was postulated as a reaction intermediate. It was assumed to take two different reaction paths: Hydrolysis was thought to give di-imine (41) – a compound unknown at that time – and 3-amino-phthalate. Unreacted diazaquinone was then thought to react with the di-imine formed to give nitrogen and luminol, the latter product formed in the excited state by this redox reaction.

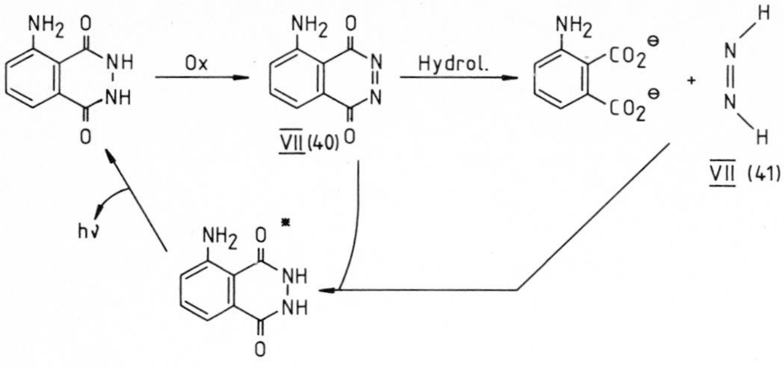

Thus, luminol itself would be the chemiluminescence emitter – an assumption which could not be adequately examined spectroscopically by the primitive apparatus then available.

In spite of the fact that this mechanism has been proved to be wrong in important details (e.g. the nature of the emitter or the role of di-imine [50]) there is experimental evidence that the diazaquinone as an intermediate may well be involved in the luminol reaction, and in luminol type chemiluminescence in general.

The kinetic experiments on luminol chemiluminescence in the system water/ alkali/potassium persulfate/hydrogen peroxide [45] were interpreted as showing that a two-electron oxidation product of luminol (i.e. the diazaquinone) was a key intermediate. The kinetics of the chemiluminescent oxidation of 7-dimethyl-amino naphthalene-1,2-dicarboxylic hydrazide led to the same conclusion [51].

Qualitatively at least luminol diazaquinone (40) can be trapped [52] in a luminol oxidation mixture (with $H_2O_2/K_3 Fe(CN)_6$) as the Diels-Alder adduct with cyclopentadiene (42).

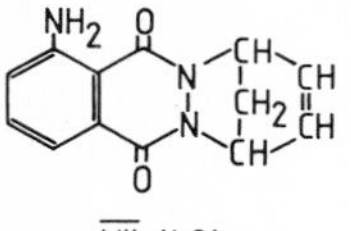

Since the preparation of the first simple diazaquinones [53, 54] (all of them being non-chemiluminescent!) chemiluminescent diazaquinones have been described. It should be pointed out that luminol diazaquinone has not been isolated as yet, nor has it been isolated from luminescent oxidation mixtures, except as the Diels-Alder adduct mentioned above.

Assuming that all chemiluminescent cyclic hydrazides show a common mechanistic path, the more easily prepared compounds to be discussed can be accepted as useful models.

By reacting the sodium salt of luminol with tert-butyl hypochlorite in dimethyl ether at $-50\,°C$, the very unstable luminol diazaquinone was obtained. It could not be purified to give an analytically pure substance, but it chemiluminesced on treatment with alkaline hydrogen peroxide with the typical blue luminol emission.

92

Benzo[b]phthalazine-1,4-dione (43) was synthesized from sodium naphthalene-2,3-dicarboxylic acid hydrazide and chlorine in diglyme at − 50°. It reacted with H₂O₂/alkali to give light with a quantum efficiency of about 5% of that of luminol and the same emission spectrum obtained from the starting hydrazide [55].

VII (43)

Since the poorly fluorescent dianion of naphthalene-2,3-dicarboxylic acid is the emitter (p. 80) the relatively low chemiluminescence efficiency of (43) is understandable.

Moreover, the reaction conditions were not optimized as there are considerable experimental difficulties. The phthalazinedione (43) is only very sparingly soluble in water which causes problems in the mixing with the oxidant. Diazaquinones are also extremely susceptible to nucleophilic attack at their carbonyl groups [54], and they have a limited lifetime in aqueous solution.

The anthracene derivative (44), naphtho-[2,3-g]-phthalazine-1,4-dione (NPD) is fairly stable, and it chemiluminesces on oxidation with basic hydrogen peroxide with an emission maximum of 430 nm, similar to the luminol emission maximum [43, 56, 57].

VII (44)

Like other diazaquinones, (44) is a deep-violet colored substance the absorption spectrum (Fig. 11) of which is characterized by a broad maximum at 520 nm. This absorption disappears slowly during decomposition in aqueous solution [56].

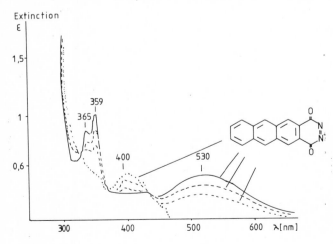

Fig. 11. Extinction spectrum of Naphthophthalazine-1.4-dione (NPD) in dimethylphthalate (——: freshly prepared solution; ...... decomposition products being formed within ca. 2h).

It also chemiluminesces on treatment with basic hydrogen peroxide. Considerably less light is obtained using oxygen or air [43]. The emission matches the fluorescence of anthracene-2,3-dicarboxylic acid (λ_{max} 430 nm). This acid was also isolated from "spent" reaction mixtures. The chemiluminescence spectrum of (44), therefore, is the same as that of the corresponding hydrazide. Neither hemin nor di-imine-forming agents/(hydroxylamine-O-sulfonate or chloroacetyl hydrazide) had any influence on this chemiluminescence. On the assumption that (44) actually is a model for the corresponding luminol derivative, the role of di-imine is thus eliminated [58]*.

Kinetic experiments, trapping the diazaquinone (44) with cyclopentadiene to give the Diels-Alder-adduct, (45), provided additional experimental proof for the role of diazaquinones in luminol type chemiluminescence. Analogous experimental evidence was provided [59] for the diazaquinone (46).

VII (45)   VII (46)

It is of course essential to connect the chemistry of the diazaquinone to hydrazide chemiluminescence if its role as an intermediate is to be confirmed. Cyclopentadiene is an excellent trap for the diazaquinone in this case as evidenced by the large rate constant for the Diels-Alder-reaction (7.7501 $M^{-1}s^{-1}$ at 21 °C). The chemiluminescence of the diazaquinone was therefore compared with that of the hydrazide in dimethylphthalate with anhydrous H_2O_2 and diethylamine as base. Heme was a necessary catalyst for the hydrazide reaction.

It was shown experimentally [60] that cyclopentadiene doesn't interfere with the use of heme as catalyst. The following partial reactions had to be taken into account:

1) Hemin (= Hem.) + DEA $\overset{k_1}{\underset{k_{-1}}{\rightleftharpoons}}$ (Hem.DEA)

2) Hem. + DEA$-H_2O_2$-complex $\overset{k_2}{\underset{k_{-2}}{\rightleftharpoons}}$ »Hem I« (Formation of the active oxidant)

3) Hem. I $\xrightarrow[k_3]{DEA-H_2O_2}$ Hem.

4) Hem. I $\xrightarrow[k_4]{DEA-H_2O_2}$ Hem.E (irreversibly destroyed hemin)

5) ADH + DEA $\overset{k_5}{\underset{k_{-5}}{\rightleftharpoons}}$ ADH$^{(-)}$ DEAH$^{(+)}$ (Formation of Hydrazide Anion)

---

* It has been reported [58] that luminol chemiluminescence is obtained when its aqueous-alkaline solutions react with hydroxylamin-O-sulfonate, provided oxygen is present. This observation could be considered as experimental evidence for the Albrecht mechanism (p. 92). If this is true, (44) should yield chemiluminescence using a di-imine-source — which does not happen.

6) $ADH^{(-)} + Hem.\ I \underset{k_{-6}}{\overset{k_6}{\rightleftharpoons}} \quad ADH_{oxid}$

7) $ADH_{oxid} \xrightarrow[\quad k'_7 \quad]{(DEA-H_2O_2)} (ADA^{2(-)})^*$ Formation of excited ADA $^{2(-)}$

8) $ADH_{oxid} \xrightarrow[\quad k'_8 \quad]{(DEA-H_2O_2)}$ By-products

9) $ADH_{oxid} + Cyclopentadiene \overset{k_9}{\rightarrow}$ Diels-Alder-Product

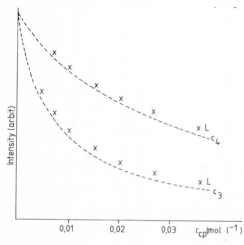

As the chemiluminescence of (44), in this homogeneous system, proceeds very rapidly (a lifetime in the millisecond range) stopped flow techniques had to be used [60].

Computer assisted calculations were carried out on the basis of the kinetic scheme shown.

Anthracene-2,3-dicar-
boxylic acid
hydrazide (ADH) $\xrightarrow{(DEA, H_2O_2, Heme)}$ NPD VII (47) $\overset{k'_{BP}}{\nearrow}$ By-products

$\xrightarrow{k'_{ADA}}$ $(ADA^{2(-)*} + ADA^{2(-)})$

$\underset{k'_{DAP}}{\searrow}$ Diels-Alder-cpd. product VII (45)

NPD: Diazaquinone, BP: by-products, ADA: anthracene-2,3-dicarboxylate. The values for k'_{BP}, k'_{ADA}, k'_{DAP} were estimated by appropriate kinetic measurements.

Fig. 12 shows good agreement, at two concentrations of basic catalyst,

Fig. 12. Measured ($\times \times \times$) and calculated ($---$) influence of cyclopentadiene concentration (CP) on the chemiluminescence light yield of (44) in the system hydrazide/H_2O_2/ diethylamin (DEA)/hemin in dimethylphthalate. DEA-concentrations: curve c_3 = 0.00446 mol l^{-1}, curve c_4 = 0.0177 mol l^{-1}.

between the calculated and observed influence of cyclopentadiene. Identical results were obtained for the diazaquinone as starting material, using only diethylamine as catalyst. The unavoidable use of an aprotic solvent may restrict the generalisation to the more usual aqueous system, but it is clear that study of diazaquinone chemiluminescence will bear directly on the mechanism of the luminol reaction.

VII.7.3 Intermediate Peroxides

The "endoperoxide" (47 a) had already been proposed as an intermediate as early as 1928. A claim had been made for the isolation of a barium salt [61]. This was later proven erroneous. It turned out to be the barium salt of luminol containing one molecule of hydrogen peroxide [62].

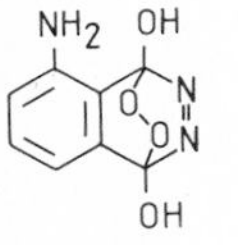

VII (47a)

The structure (47) for the intermediate is attractive, both in its likely method of formation, and in the way in which it accounts neatly for the products and distribution of $^{18}O_2$. It has not yet been isolated, and is likely to be exceedingly unstable if formed. It could be generated in a variety of ways, depending upon the reaction conditions for luminol chemiluminescence.

Nucleophilic attack on one of the diazaquinone carbonyls by hydrogen peroxide monoanion, followed by deprotonation of the hydroperoxide and ring closure is an attractive possibility.

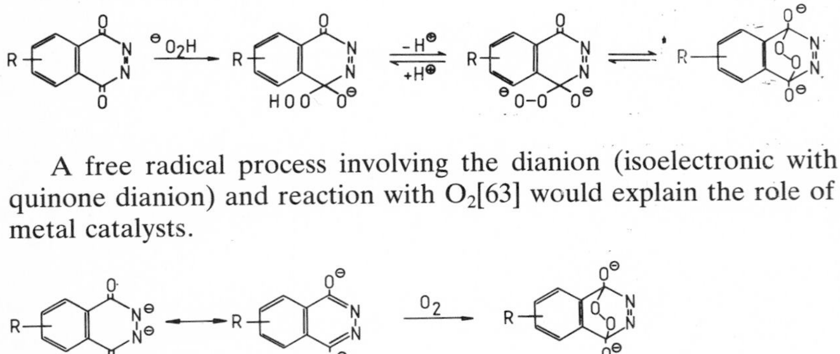

A free radical process involving the dianion (isoelectronic with a hydro-quinone dianion) and reaction with O_2[63] would explain the role of transition metal catalysts.

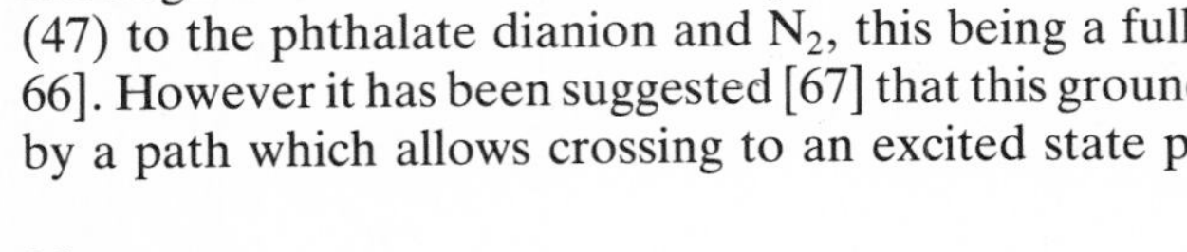

It has been suggested that this reaction with oxygen may be reversible [64], although this seems rather unlikely. There is no barrier to the cycloreversion of (47) to the phthalate dianion and N_2, this being a fully allowed process [47, 65, 66]. However it has been suggested [67] that this ground state process can proceed by a path which allows crossing to an excited state potential energy surface.

Several alternative mechanisms have been proposed, differing in detail but similar in general; nitrogen elimination is their common step.

A retro-Diels-Alder reaction would lead to the formation of the xylene peroxide. According to Michl [67, 68] this should be a reaction requiring an activation energy of only 63 kJ/mol.

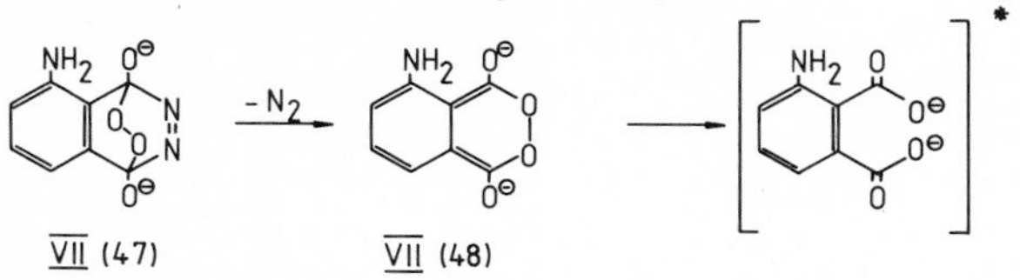

VII (47)   VII (48)

Decomposition of (48) to yield excited 3-amino phthalate is then expected since there is a path from the 18π-electron ground state of (48) to the $n,n \rightarrow \pi^*,\pi^*$ state of the aminophthalate [67].

It is however difficult to agree with this since azocompounds of this sort usually require irradiation or elevated temperatures for N_2 expulsion. Thus (49) forms o-xylylene on irradiation at $-196\,°C$ in rigid media [68], with formation of (50) and (51) on slight warming. It is stable at room temperature in the dark.

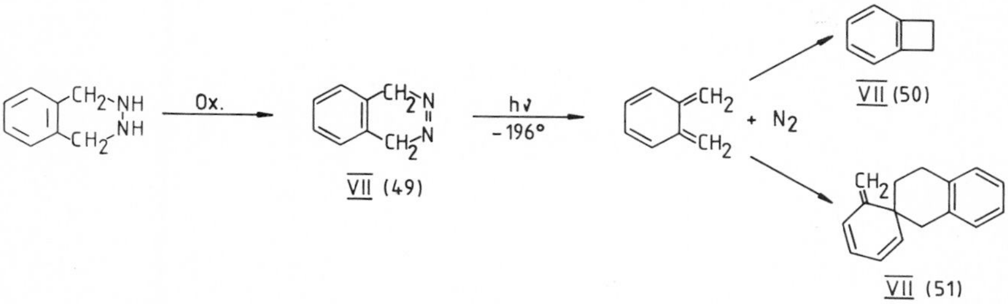

VII (50)

VII (49)

VII (51)

A good analogy for the proposed reaction of the endoperoxide (47) is the chemiluminescent thermolysis of the pyranone endoperoxide (52) in the presence of appropriate fluorescers [69, 70].

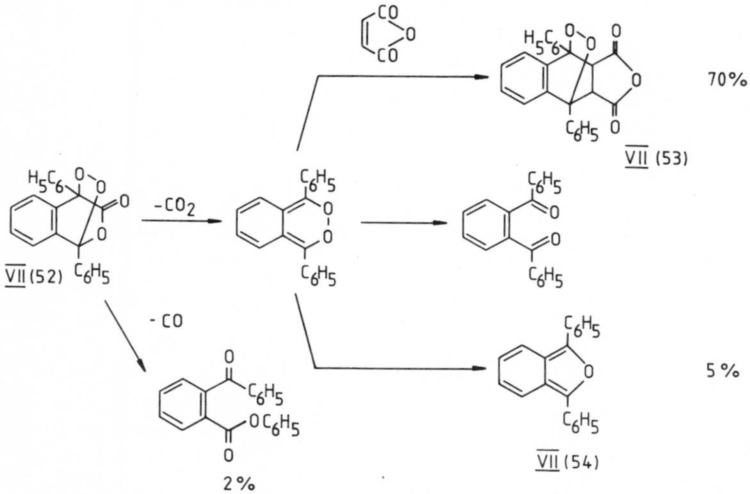

70%

VII (53)

VII (52)

5%

VII (54)

2%

97

In this case, an o-xylylene peroxide is formed, confirmed by the isolation of the Diels-Alder adduct (53) with maleic anhydride in 70% yield, and by infrared spectroscopy. If the peroxide (52) is heated to ca. 100°, o-dibenzoylbenzene is formed in 85% yield, together with ca. 5% of 1,3-diphenyl isobenzofuran (54) and 2% phenyl o-benzoyl benzoate (55).

Some chemiluminescence accompanies this reaction, and emission matches the fluorescence of the isobenzofuran derivative (54), presumably excited by energy transfer or direct by CIEEL.

A much stronger chemiluminescence is observed when fluorescent compounds such as perylene, DPA or 7-(dimethylamino)-2-methylphenazine are present. As the chemiluminescence intensity varies linearly with the one-electron oxidation potential of the activator, the CIEEL mechanism (see p. 34) is suggested as operating in this case [71].

This chemiluminescent reaction of (52) can serve as model for the chemiluminescent decomposition of the "luminol endoperoxide" [67] the only problem being at present that evidence for the luminol endoperoxide is so hard to obtain.

If the CIEEL mechanism is adopted, the intramolecular character of this particular electron exchange mechanism has to be considered. The amino group of luminol – or other electron-rich groups in luminol type hydrazides – transfers one electron to the peroxide grouping:

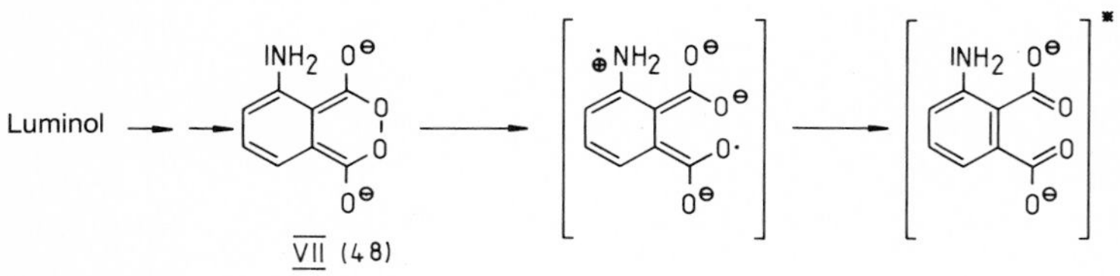

$$\overline{\text{VII}} \ (48)$$

Another pathway for nitrogen elimination has also been formulated [72–74]. The peroxycarboxylic aldehyde (56) formed in the first step, equilibrates with a cyclic peroxy semiacetal (103). An intramolecular CIEEL was proposed [73].

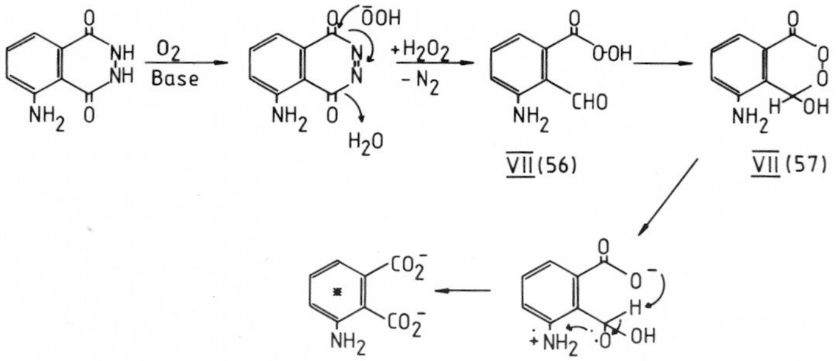

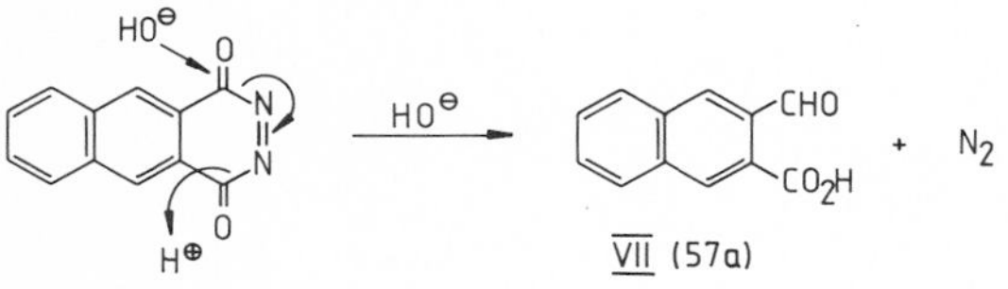

$$\overline{\text{VII}} \ (56) \qquad \overline{\text{VII}} \ (57)$$

$$\overline{\text{VII}} \ (57a)$$

98

This mechanism (or a variant) is attractive in that it follows a similar path to that which produces the *ortho*-carboxyaldehyde (57 a) with $HO^{(-)}$ rather than $HO_2^{(-)}$ as nucleophile [73]. The intramolecular electron transfer is in fact a combination of the hypothesis for the firefly dioxetanone decomposition (p. 152) and the (intramolecular) chemiluminescent activation of secondary peresters (p. 35). However it is only one hypothesis among many which lack confirmatory evidence in this exceedingly complicated reaction.

VII.7.4 1,4-Dialkoxy Phthalazines as Potential Precursors of the Luminol-"endoperoxide"

An interesting alternative approach to the synthesis of the endoperoxide produced a highly significant negative result.

A series of substituted 1,4-dimethoxy phthalazines (58) were treated with singlet oxygen, produced by photosensitisation [75]. The corresponding phthalates were formed smoothly by elimination of N_2, the endoperoxides being the assumed intermediates:

R = H , OCH_3 or NH_2

The methoxy compound ($R = OCH_3$) yielded the strongly fluorescing 3-methoxy phthalate. No chemiluminescence was observed, however, when (58) was treated with singlet oxygen in methanolic solution at $-78°$ and the reaction product warmed to room temperature.

VII.7.5 Luminol Radical Anions as Intermediates (Pulse Radiolysis Experiments)

As mentioned above, luminol (and other cyclic hydrazide) radical anions are produced by a wide range of reagents.

Pulse radiolysis often gives useful insight into difficult mechanisms, providing information on transient intermediates and simplifying the conditions. However in its application to the study of luminol, there is difficulty in connecting the observations with those in the more usual systems.

Nevertheless Baxendale's pioneering work [76] must certainly be kept in mind during discussions of more usual luminol systems. Initiation occurred by the HO· radical, with oxygen and hydroxide ion the only other requirements. Transients absorbing in the range 220–1000 nm were detectable and the decay rate of 100 µs indicated that they were connected with chemical events rather than excited states. A notable feature in this work, not considered in other explanations of the mechanism is the second order nature of the excitation reaction. The scheme shown was devised to explain this.

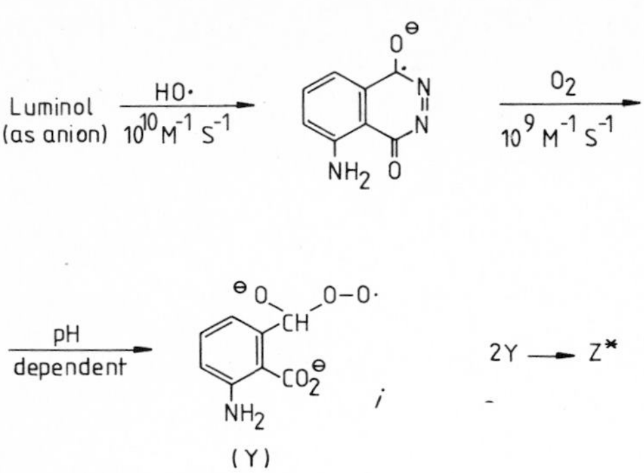

The last dimerisation reaction is probably assumed to resemble the Russell termination step (p. 23) but it must be said that is an unlikely candidate for the typically high quantum yields of the luminol reaction. Direct reaction [76] with oxygen is observed and, instead of diazaquinone, its addition product with oxygen radical anion is assumed. An endoperoxide such as (47) is not proposed as an intermediate.

A similar study has been carried out by the pulse radiolytic production of luminol radical anions and oxygen radical ions simultaneously [77].

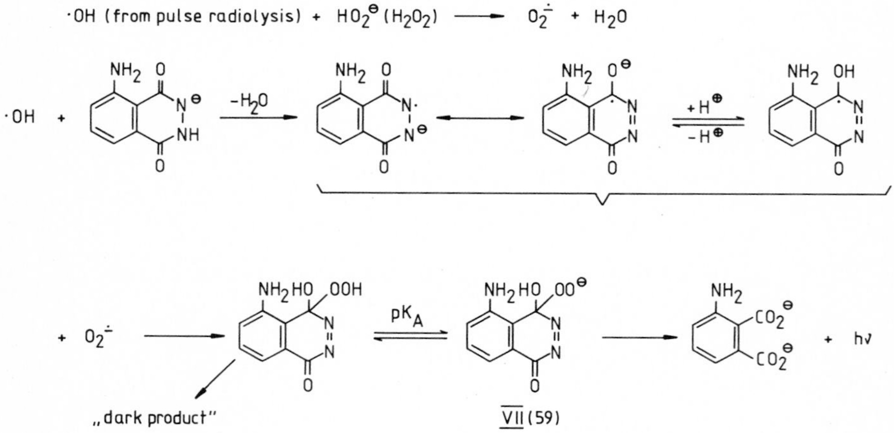

An aqueous-alkaline solution containing hydrogen peroxide, but no molecular oxygen was used. The rate of the luminol peroxide formation was assayed at different pH values.

Since there is no effect of H_2O_2 concentration on the integrated chemiluminescence intensity, the conclusion was drawn that oxygen radical anion "activates" the luminol radical anions to chemiluminescence, forming the non-radical "diazaquinone perhydrate" (59):

The rate constants for the formation of the peroxide are $15.6 \times 10^{-8}M^{-1}s^{-1}$ (at pH 7.7) and $1.6 \times 10^{-8}M^{-1}s^{-1}$ (at pH 11).

From the transient absorption spectra of the pulsed solutions of luminol and hydrogen peroxide, with the appearance of a band at 500 nm, it is assumed that a small amount of diazaquinone is present [77].

Again a rather strong dependence on pH was observed for the *decomposition* of the luminol-oxygen adduct with

k at pH 9.3 = $1.71–3.67 \times 10^{-3}\text{s}^{-1}$ and
k at pH 11.0 = $200 \times 10^{-3}\text{s}^{-1}$

A similar pH dependence has been observed before [8] for the oxidation of luminol in the classical reaction. It is not known whether the steps involved are the same.

The "dark product" in scheme p. 134 is as yet unknown. It is assumed to be formed from the protonated peroxide.

The non-protonated form leads to chemiluminescence and $\varnothing_{CL}$ is estimated to be nearly 0.10 which is about ten times greater than that of luminol (p. 91) in the hemin-catalyzed aqueous system.

A comparison with the results obtained in a luminol/O_2 system [76, 78] does however reveal some parallels. Among these are the absorption between 500 and 600 nm pointing to a diazaquinone intermediate [78].

These results involving the luminol radical anion are compatible with other proposals for the luminol mechanism. Steps after the peroxide intermediate are not specifically detected, and must occur with a rate constant of 10^8 s^{-1}.

VII.7.6 Present State of Hypotheses on the Luminol Chemiluminescence Mechanism

The following scheme represents a synopsis of the probable pathways of the luminol chemiluminescence mechanism:
a) In water and other solvents:
luminol → luminol-monoanion → luminol-radical-anion → peroxide (59 a)

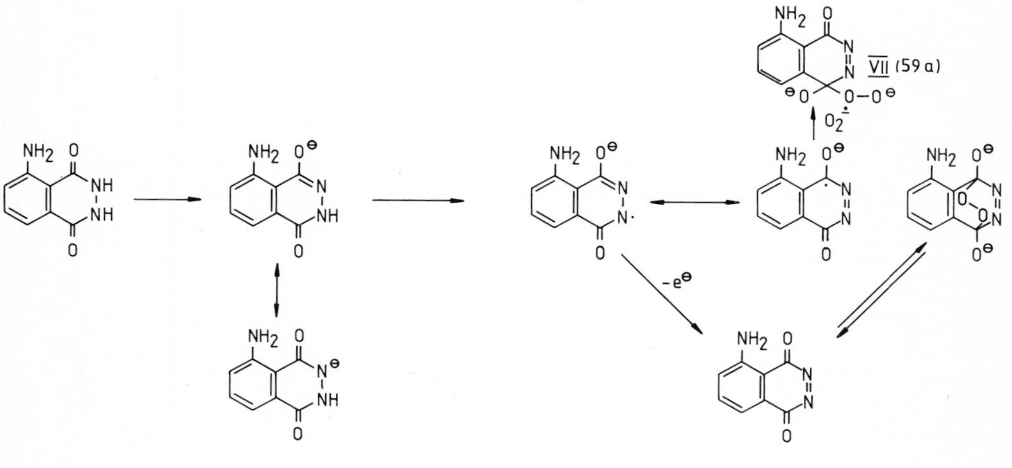

Whether the peroxide is a radical or an anion, should depend on the oxidative species reacting with the luminol radical anion. The peroxide (59 a) could also be formed, as was mentioned before (p. 96), from the diazaquinone (40) and oxygen or oxygen radical anion. The peroxide (59 a) might undergo cyclisation to the "endoperoxide" (47), a seemingly inescapable intermediate in any sequence of oxidations. Whether this confirms it as an obligate intermediate in the route to excitation is less clear.

b) In aprotic solvent (DMSO etc.):

luminol → luminol dianion → diazaquinone → peroxide → endoperoxide (47)

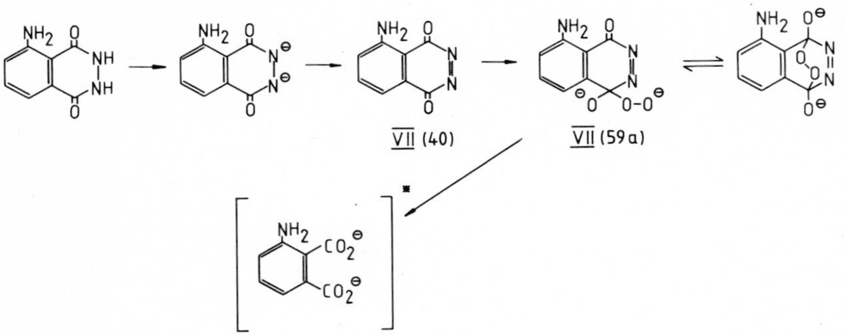

Under both sets of conditions the diazaquinone is implicated but it is quite possible that it is a by-product, albeit a chemiluminescent one. A reasonable explanation of the need for catalysts in the aqueous system is the greater difficulty in oxidizing the mono-anion (either to the radical or through to the diaza-quinone).

The more easily oxidized dianion, possible only in dipolar aprotic solvents, succumbs to O_2 alone. The reaction of the radical (or radical anion) with oxygen may lead to the endoperoxide, and thus join the pathway from the diazaquinone. Other peroxidic products leading to chemiluminescence have not been excluded.

The decomposition of the endoperoxide via the o-xylene derivative (48) in an intramolecular electron transfer mechanism would also give the substituents a decisive role in the excitation step itself – not only affecting the fluorescence efficiency of the phthalate dianion. A high fluorescence efficiency is of course a necessary, but not sufficient, requirement in luminol type chemiluminescence [9, 16].

The results of White and coworkers [9, 16], concerning the excitation yields of luminol and isoluminol derivatives demonstrate that a simple correlation be-tween oxidation potential of the substituent group and chemiluminescence efficiency (as in many CIEEL examples, see p. 40) is not observed in the luminol case.

Not all of the factors responsible for chemiluminescence will respond in the same way to changes in substitution. However even when separately determined for alkyl substituted isoluminol derivatives (Table 9), the relatively small differ-ences observed are difficult to interpret. The fluorescence quantum yield stays reasonably constant but the efficiency of excited state population ($\emptyset_{ES}$) shows no clear trend.

Table 9. (after Brundrett, Roswell and E. H. White [9])

Compound	$\varnothing$ Chemi-luminescence[1]	$\varnothing$ Fluorescence of the corresponding acids[2] (in 0.1 M K_2CO_3)		$\varnothing_{ES}$ (efficiency of excited product formation)
I	0.000047	0.029	(V)	0.0017
II	0.00052	0.12	(VI)	0.0043
III a	0.0012	0.13	(VII a)	0.0092
III b	0.0075	0.25	(VII b)	0.030
III e	0.0094	0.31	(VII c)	0.030
III d	0.0140	0.30	(VII d)	0.046
III e	0.0038	0.26	(VII e)	0.015
III f	0.0058	0.28	(VII f)	0.021
IV a	0.015	0.30	(VIII a)	0.050
IV b	0.028	0.26	(VIII b)	0.10

[1] Obtained in aqueous 0.1 M K_3CO_3 solution ($p_H = 11.4$) with hydrogen peroxide and hemin; the values are relative to luminol (0.0125 according to Lee and Seliger [91]). The precision of the measurements is about $\pm 10\%$.
[2] Values relative to quinine bisulfate 0.55. It should be mentioned that even small changes in reaction conditions (*e. g.* change of pH) give rise to changes in $\varnothing_{CL}$.

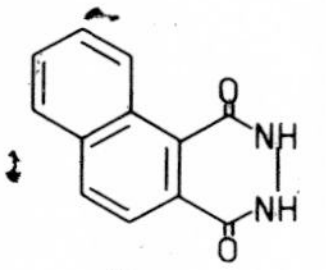

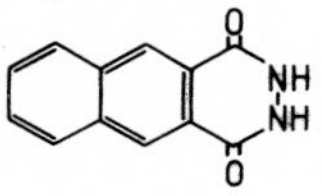

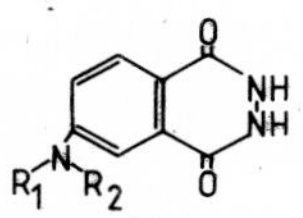

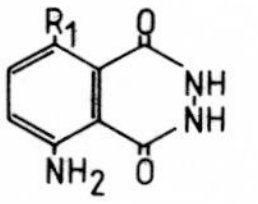

a: $R_1 = R_2 = H$
b: $R_1 = R_2 = CH_3$
c: $R_1 = R_2 = C_2H_5$
d: $R_1 = R_2 = n-C_4H_9$
e: $R_1 = R_2 = n-C_7H_{15}$
f: $R_1 = H$; $R_2 = C_4H_9$

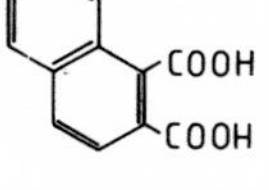

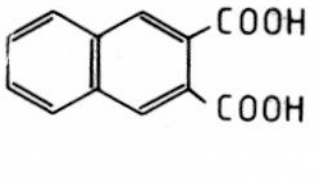

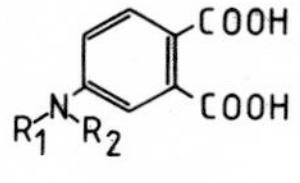

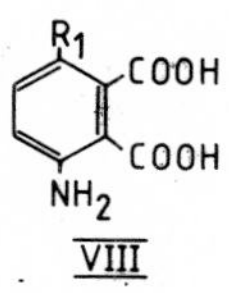

The more substantial changes in substitution do show a relationship with ionisation potential, interpreted as being consistent with an electron transfer mechanism [9]. The factors $\varnothing_F$, $\varnothing_{ES}$ and $\varnothing_{CL}$ were determined in the usual way.

For example $\varnothing_{CL}$ increase from I to IV b by a factor of 600, in line with electron donation.

Fluorescence ($\varnothing_F$) also increases (by a factor of 10). Thus $\varnothing_{ES}$ increases by 60 fold over the series, and the increase correlates quite well with the decreasing energy of the excited state as measured by the O-O band of the more accessible υ_{max}. As discussed in electron transfer chemiluminescence, this is usually taken to indicate that an electron is transferred in the excitation step, although in this case the step itself is unidentified.

VII.8 Monoacylhydrazides

With a few exceptions, monoacyl hydrazides chemiluminesce with efficiencies some 100 fold lower than the cyclic diacylhydrazides. With the assumption that the mechanism of the luminol type chemiluminescence involves an o-xylene peroxide (p. 97), this behaviour of the linear hydrazides would be easily understood: such an intermediate is not possible in the linear case.

Although the efficiency of the monoacyl hydrazides is generally low, there are exceptions. These few cases can in fact approach the efficiency of luminol. There is little doubt that the mechanism is different, but here again it is not possible to write a convincing general scheme.

However the same catalysts and solvent systems are effective, and molecular oxygen is involved. A common feature in those most studied is the apparent involvement of the acyl *anion* and the formation of the carboxylate in the excited state.

The more efficient examples are shown below:

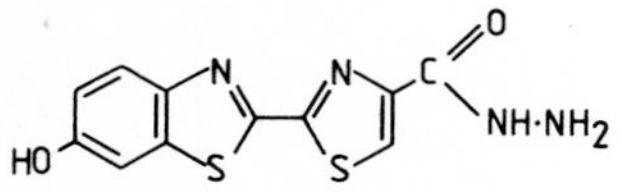

$\overline{VII}$ (60)

$\varnothing_{CL}$: 5 × 10^{-4} [80]

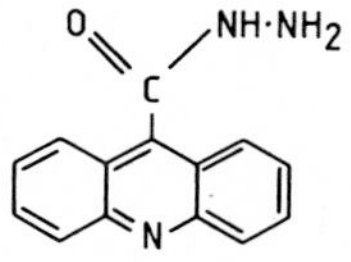

$\overline{VII}$ (61)

$\varnothing_{CL}$: 3 × 10^{-4} [80]

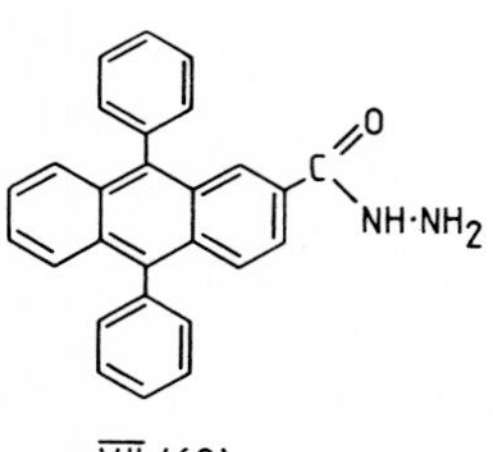

$\overline{VII}$ (62)

$\varnothing_{CL}$: ca. 10^{-6} [81]

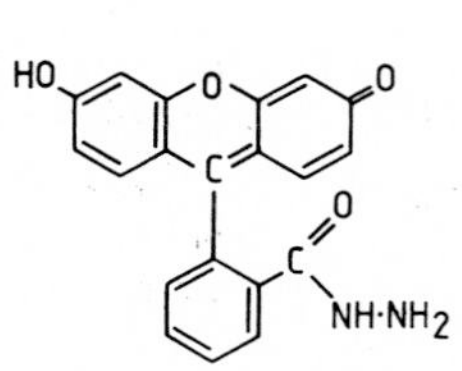

$\overline{VII}$ (63) $\varnothing_{CL}$: 6 × 10^{-5} [82]

(All in alkaline DMSO/O$_2$)

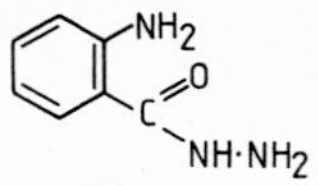

VII (64)

(H₂O/alkali/H₂O₂)
$\emptyset_{CL}$: ca. 10^{-7} [82]

VII.8.1 Chemiluminescence Mechanism of Monoacylhydrazides

Although direct evidence is lacking, White [80] has made a very good case for the intermediacy of the acyl anion. In contrast to the cyclic hydrazides, substitution on the nitrogen (of the hydrazide group) does not seriously affect chemiluminescence.

A mechanism can be written which applies to any hydrazide with free NH-groups:

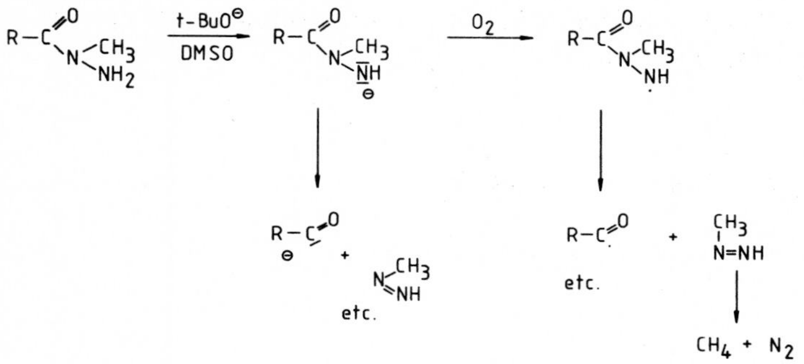

Much of the evidence for this route came from the initially surprising oberservation that acridine 9-carboxylic acid hydrazide gave not the excited carboxylate, but emission from acridone. Strong support for the acyl anion route is obtained from the fact that 9-formyl acridine in strong base is highly chemiluminescent [80].

The formation of excited acridone can be explained by one of three mechanisms:

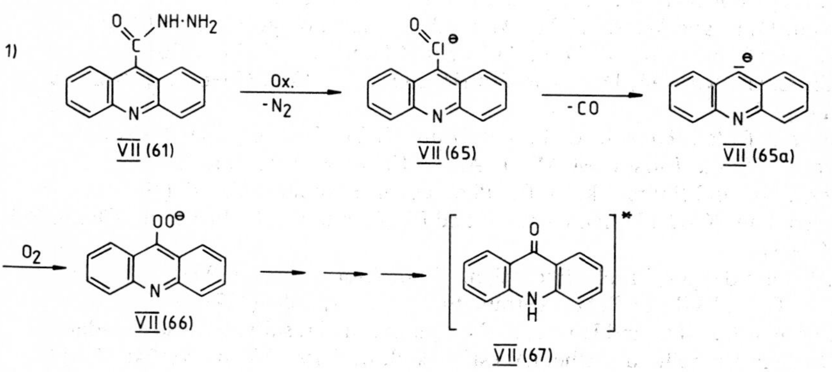

1)

VII (61)   VII (65)   VII (65a)

VII (66)   VII (67)

105

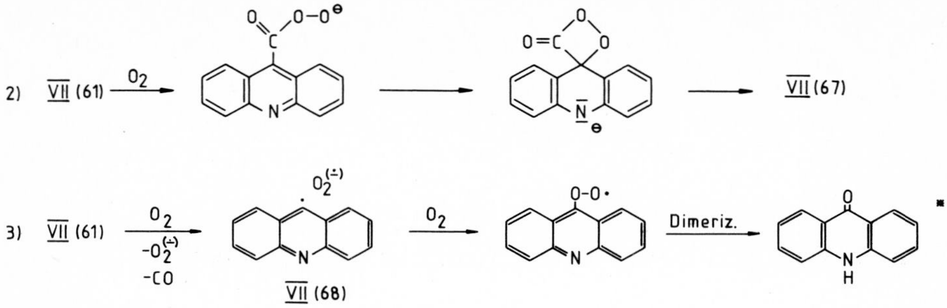

2) $\underline{\text{VII}}$ (61) $\xrightarrow{O_2}$ $\longrightarrow$ $\longrightarrow$ $\underline{\text{VII}}$ (67)

3) $\underline{\text{VII}}$ (61) $\xrightarrow[\substack{-O_2^{(-)} \\ -CO}]{O_2}$ $\xrightarrow{O_2}$ $\xrightarrow{\text{Dimeriz.}}$

$\underline{\text{VII}}$ (68)

The intermediate 9-acridinyl peroxide may react by dimerisation or by a route related to that of the Grignard chemiluminescence (p. 28).

That the chemiluminescence of the monoacyl hydrazides follows mechanisms other than that of the luminol type hydrazides, is evident from the fact that methyl substitution in the hydrazide group does not lead, as with luminol to the loss of chemiluminescence. The hydrazides (69) and (70), for example, chemiluminesce on oxidation in the DMSO system:

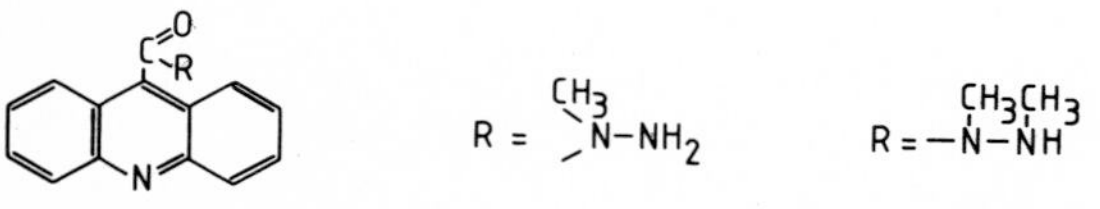

$\underline{\text{VII}}$ (69)

$R = \overset{CH_3}{\underset{}{>}}N-NH_2$

$R = -\overset{CH_3}{\underset{}{N}}-\overset{CH_3}{\underset{}{N}}H$

$\underline{\text{VII}}$ (70)

VII.9 References

1. Gundermann, K.-D., Chemilumineszenz organischer Verbindungen, p. 63, Springer-Verlag, Berlin 1968
2. Gundermann, K.-D., Topics Curr. Chem. *46,* 61 (1974)
3. White, E. H., Miano, J. D., Watkins, C. J. and Breaux, E. J., Angew. Chem. *86,* 292 (1974), Int. Ed. Engl.
4. McCapra, F., in: Bradley, J. N., Gilliard, R. D. and Hudson, R. F. (Eds.) Essays in Chemistry, Vol. *3,* p. 10, Academic Press New York 1972
5. Roswell, D. F. and White, E. H., Methods in Enzymology *57,* 409 (1978)
6. White, E. H. and Roswell, D. F., Accounts Chem. Res. *3,* 54 (1970)
7. White, E. H., Roswell, D. F. and Zafiriou, O. C., J. Org. Chem. *34,* 2462 (1969)
8. Lit. 1), p. 71
9. Brundrett, R. B., Roswell, D. F. and White, E. H., ibid. *94,* 7536 (1972)
10. White, E. H. and Bursey, M. M., J. Amer. Chem. Soc. *86,* 941 (1964)
11. Totter, J. R. and Philbrook, G. E., Photochem. Photobiol. *5,* 177 (1966)
12. Gundermann, K.-D., Horstmann, W. and Bergmann, G., Liebigs Ann. Chem. *684,* 127 (1965)
13. Lee, J., and Seliger, H. H., Photochem. Photobiol. *4,* 1015 (1965)
14. Wei, C. C. and White, E. H., Tetrahedron Lett. *31,* 3351 (1971)
15. Gundermann, K.-D., and Röker, K.-D., unpublished results; see also Gundermann, K.-D., Schriftenreihe der Rhein.Westf. Akad. d. Wiss., Vortr. Nr. 246 (1975)

16. Lee, J. and Seliger, H. H., Photochem. Photobiol. *11,* 247 (1970)
17. Gundermann, K.-D., Preprints of the International Symposium on Chemiluminescence, Durham, N. C., p. 312 (1965)
18. Gundermann, K.-D., Chimia *25,* 261 (1971)
19. McCapra, F., Pure Appl. Chem. *24,* 611 (1970)
20. Brundrett, R. B. and White, E. H., J. Amer. Chem. Soc. *96,* 7497 (1974)
21. Drew, H. D. K. and Pearman, F. H., J. Chem. Soc. (London) *1937,* 586
22 a. Gundermann, K.-D. and Kaufung, R., unpublished results;
22 b. R. Kaufung, Diplomarbeit, TU Clausthal 1981
23 a. Gundermann, K.-D. and Nordmann, J., unpublished results;
23 b. J. Nordmann, Diplomarbeit, TU Clausthal 1981
24. Spengler, H., Dissertation, TU Clausthal 1969
25. Lüpke, U., Diplomarbeit, TU Clausthal
26. Gundermann, K.-D. and Giesecke, H., Liebigs Ann. Chem. *1979,* 1985
27. Haase, B., Diplomarbeit, TU Clausthal 1979
28. White, E. H., Roswell, D. F., Wei, C. C. and Wildes, P. D., J. Amer. Chem. Soc. *94,* 6223 (1972)
29. Gorsuch, J. D. and Hercules, D. M., ibid. *15,* 227 (1972)
30. Lee, J. and Seliger, H. H., Photochem. Photobiol. *15,* 567 (1972)
31. Bersis, D. and Nikokavouras, J., Nature (London) *217,* 451 (1968)
32. Omote, Y., Yamamoto, H. and Sugiyama, M., J. Chem. Soc., Chem. Commun. *1976,* 914
33. Lee, J. and Seliger, H. H., Photochem. Photobiol. *15,* 227 (1972)
34. Gundermann, K.-D. and Drawert, M., Chem. Ber. *95,* 2018 (1962)
35. Lee, J. and Seliger, H. H., Photochem. Photobiol. *11,* 247 (1970)
36. White, E. H. and Roswell, D. F., J. Amer. Chem. Soc. *89,* 3944 (1967)
37. Roswell, D. F., Paul, V. and White, E. H., ibid. *92,* 4855 (1970)
38. Roberts, D. R. and White, E. H., ibid. *92,* 4861 (1970)
39. Gundermann, K.-D. and Röker, K.-D., Liebigs Ann. Chem. *1976,* 140
40. Gundermann, K.-D., Angew. Chem. *85,* 451 (1973); Int. Ed. Engl. *12,* 425 (1973)
41. Röker, K.-D., Dissertation, TU Clausthal 1974
42 a. Brinkmeyer, H., Dissertation, TU Clausthal 1978;
42 b. Gundermann, K.-D., in: DeLuca, M. A. and McElroy, W. D., (Eds.), Bioluminescence and Chemiluminescence p. 17, Academic Press, New York, 1981
43. Gundermann, K.-D., Fiege, H. and Klockenbring, G., Liebigs Ann. Chem. *738,* 140 (1970)
43 a. White, E. H. and Bursey, M. M., J. Org. Chem. *31,* 1912 (1966)
44. Gundermann, K.-D. and Schedlitzky, D., Chem. Ber. *102,* 3241 (1969)
45. Rauhut, M. M., Semsel, A. M. and Roberts, B. G., J. Org. Chem. *31,* 2431 (1966)
47. White, E. H. and Brundrett, R. B., in: Cormier, M. J., Hercules, D. M. and Lee, J., (Eds) Chemiluminescence and Bioluminescence, p. 231, Plenum Press, New York 1973
48. White, F. H., Zafiriou, O. C., Kaegi, H. H. and Hill, J. H. M., J. Amer. Chem. Soc. *86,* 940 (1964)
48 a. Brundrett, R. B. and White, E. H., ibid. *96,* 7497 (1974)
49. Albrecht, H. O., Z. phys. Chem. *136,* 321 (1928)
50. Lit. 1), p. 83
51. Eicke, H. F., Fiege, H. and Gundermann, K.-D., Z. Naturforsch. *25 b,* 484 (1970)
52. Omote, Y., Miyake, T. and Sugiyama, N., Bull. Chem. Soc. Japan *40,* 2446 (1967)
53. Clement, R. A., J. Org. Chem. *25,* 1724 (1960)
54. Kealy, T. J., J. Amer. Chem. Soc. *84,* 966 (1962)

55. White, E. H., Nash, E. G., Roberts, D. R. and Zafiriou, O. C., J. Amer. Chem. Soc. *90*, 5932 (1968)
56. Gundermann, K.-D., in Cormier, M. J., Hercules, D. M. and Lee, J. (Eds.) Chemiluminescence and Bioluminescence, p. 209, Plenum Press, New York 1973
57. Gundermann, K.-D. and Fiege, H., Liebigs Ann. Chem. *743*, 200 (1971)
58. cf. Metcalf, W. S. and Quickenden, T. I., Nature (London) *206*, 507 (1965)
59. Rusin, A., Leksin, N. A. and Roshshin, A. L., J. Org. Ch. (russ.), Engl. Translat. 199 (1980)
60. Gundermann, K.-D., Unger, H. and Stauff, J., J. Chem. Res. (S) *1978*, 318; (M) *1978*, 3846
61. Drew, H. D. K. and Garwood, R. F., J. Chem. Soc., 791 (1938)
62. White, E. H., in: McElroy, W. D. and Glass, B., A Symposium on Light and Life, p. 183, The Johns Hopkins Press, Baltimore 1961
63. Seitz, W. R., J. Phys. Chem. *79*, 101 (1975)
64. Beck, M. T., Joo, F., Photochem. Photobiol. *16*, 491 (1972)
65. Unger, H., Diplomarbeit, TU Clausthal 1971
66. Michl, J., in: Eyring, H., Henderson, D. and Jost, W. (Eds.), Physical Chemistry VII, p. 125, Academic Press New York 1975
67. Michl, J., Photochem. Photobiol. *25*, 141 (1977)
68. Flynn, C. R. and Michl, J., J. Amer. Chem. Soc. *96*, 3280 (1974)
69. Smith, J. P. and Schuster, G. B., ibid. *100*, 2546 (1978)
70. Smith, J. P., Schrock, A. K. and Schuster, G. B., ibid. *104*, 1041 (1982)
71. Adam, W. and Erden, I., ibid. *101*, 5692 (1979)
72. McCapra, F., J. Chem. Soc., Chem. Commun. *1977*, 946
73. McCapra, F., Internat. Conference on Chemi- and Bio-Energized Processes, Guarujà-Sao Paulo, 1978: see Gundermann, K.-D., in: Schram, E. and Stanley, P. (Eds.), Internat. Symposium on Analytical Application of Bioluminescence and Chemiluminescence, p. 72, State Printing & Publishing Inc. Westlake (Calif.) 1979
74. McCapra, F. and Leeson, P. D., J. Chem. Soc., Chem. Commun. *1979*, 114
75. Goto, T., Isobe, M. and Ienaga, K., in: Cormier, M. J., Hercules, D. M. and Lee, J. (Eds.), Chemiluminescence and Bioluminescence, p. 492, Plenum Press, New York 1973
76. Baxendale, J. H., J. Chem. Soc., Chem. Commun. *1971*, 1489
77. Merényi, G. and Lind, J. S., J. Amer. Chem. Soc. *102*, 5830 (1980)
78. Würzburg, F. and Haas, Y., Chem. Phys. Lett. *55*, 250 (1978)
79. White, E. H. and Roswell, D. F., in: Chemi- and Bioluminescence (Burr, J., ed.), p. 215, Dekker, N. Y. 1985
80. Rapaport, E., Cass, M. W. and White, F. H., J. Amer. Chem. Soc. *94*, 3153 (1972)
81. Gundermann, K.-D. and Brinkmeyer, H., unpublished results; Brinkmeyer, H., Diplomarbeit TU Clausthal 1975
82. Nikokavouras, J., Zois, J., Vassilopoulos, G. and Perry, A., J. prakt. Chem. *323*, 21 (1981)
83. Ojima, H., Naturwiss. *48*, 600 (1961)
84. Gundermann, K.-D., Lathia, D., Nolte, W. and Röker, K.-D., Liebigs Ann. Chem. *1974*, 798
85. Gundermann, K.-D., Röker, K.-D., Klockenbring, G. and Brinkmeyer, H., ibid. *1976*, 1873

VIII. Acridine Derivatives

Only a few years after the first paper on luminol, Gleu and Petsch [1], in 1935, published the first report on lucigenin (2) chemiluminescence

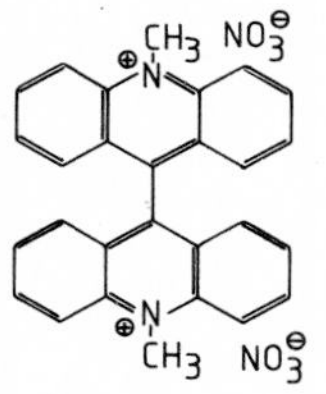

$\underline{\text{VIII}}$(2)

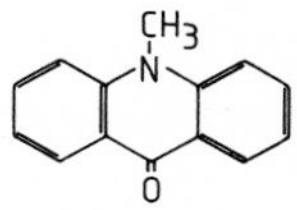

$\underline{\text{VIII}}$(3)
= (NMA)*

It was eventually shown that the emitting species is N-methyl acridone (3) – provided that at the appropriately low concentration, no energy transfer to secondary reaction products, formed from (3) in the aqueous alkaline medium, takes place. N-methyl acridone fluoresces at λ_{max} 445 nm, that is in the blue region of the spectrum. That greenish chemiluminescence is generally observed is due to energy transfer from (3) to lucigenin, to dimethyl biacridylidene and other compounds (for details see [2, 18]).

N-methyl acridone is considerably more fluorescent than 3-amino phthalate, the emitting species in the luminol reaction (see Chap. VII), and so are numerous other acridine derivatives.

Therefore many efforts have been made to take advantage of the chemiluminescence of lucigenin and acridine derivatives, especially for analytical purposes.

VIII.1 Lucigenin

VIII.1.1 The Reaction Mechanism

As early as in the first paper on lucigenin chemiluminescence [1] the dioxetan derivative (1) had been postulated as a key intermediate:

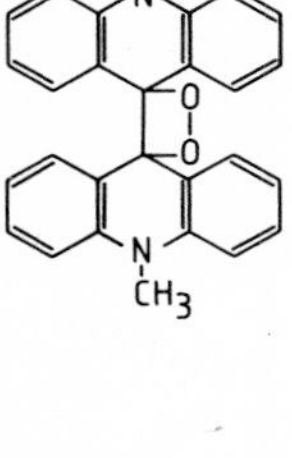

$\underline{\text{VIII}}$ (1)

However (1) was thought to be formed by dehydrogenation of the lucigenin pseudo base by hydrogen peroxide, an extremely unlikely reaction. This redox reaction was thought to be the actual cause of chemiluminescence. That simple mechanism cannot be correct for 2 reasons: a) the energetic requirements are not met, as the proposed redox reaction provides ca. 189 kJ/mol only, whereas the observed green emission requires about 252 kJ/mol.

b) N-methyl acridone, not known at that time to be the primary excited product, is not part of the mechanism.

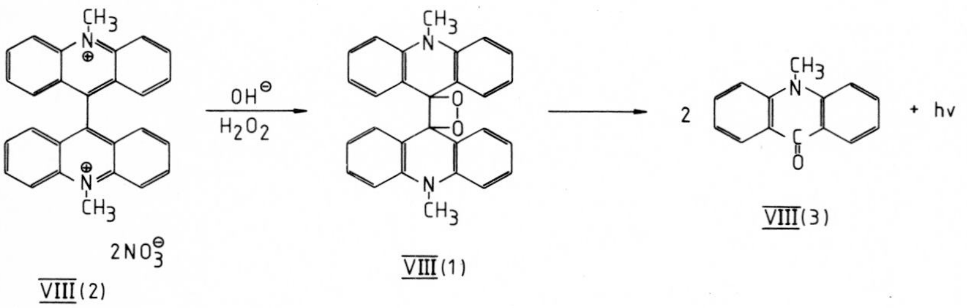

Nonetheless, a dioxetan decomposition mechanism for lucigenin chemiluminescence, based on the exergonic processes described in Chap. V, seems well established [3]. A direct demonstration of the intermediacy of this dioxetane was first made [4] in 1969 by treating 10,10'-dimethyl-9,9'-biacrylidene (4) with singlet oxygen from several sources. Emission from N-methyl acridone was unequivocally shown. The lifetime of the intermediate was characteristic of the supposed dioxetane. Intramolecular electron transfer has been suggested as the excitation mechanism in the decomposition of this and other electron-rich dioxetans.

A more detailed study [5] confirmed these observations

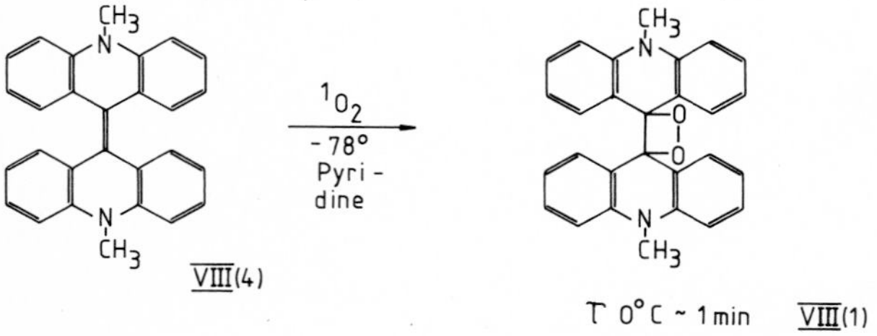

and the Arrhenius parameter obtained supported the structure of the intermediate. The primary excited product–provided concentrations of 10,10'-dimethyl-9,9'-biacrylidylidene (4) smaller than 10^{-4} mol are used – was again shown to be N-methyl acridone (3) (Fig. 13).

If the concentrations of (2) in the solutions are higher, singlet-singlet energy transfer from (3) to (2) takes place. This can be seen in Fig. 13 (from [5]):

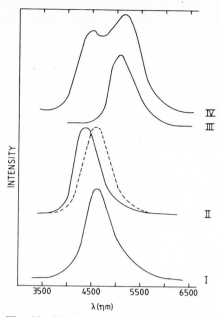

Fig. 13. Chemiluminescence during warming up after reaction of 10,10′-dimethyl-9,9′-biacrylidylidene with $^1O_2$.
(The influence of zinc tetraphenylporphin on the emission spectrum is taken into account, as ZnTPP was used as sensitizer in the singlet oxygen reaction; it causes a blue shift by internal absorption).

   I: Normalized Fluorescence Spectrum of *(3)*
  II: in the Absence (——) and Presence (– – – –) of ZnTPP (10^{-4} M)
 III: Fluorescence Spectrum of (4) in the Presence of ZnTPP (10^{-4} M)
 IV: Chemiluminescence from incomplete Reaction of (4) with singlet oxygen.

Dioxetan formation is thought to occur according to the following mechanism:

111

A very thorough examination [9] of this mechanism has added useful corroborative detail. The salient features described by these workers are that the reaction is first order in lucigenin with the rate expression (L = lucigenin):

$$\frac{d\,[L]}{dt} = [L]\,k_1\,[HO_2^{(-)}] + k_2\,[HO_2^{(-)}]\,[HO^{(-)}]$$

Use of tert-butyl hydroperoxide instead of H_2O_2 produced a slower reaction (although lucigenin disappeared faster). With H_2O_2, $\varnothing_{CL}$ is comparable to that of luminol (1.25×10^{-2}) but it falls to about 100 fold less with the alkyl hydroperoxide as oxidant.

Aerobic oxidation was responsible at least in part for this weak light. The influence of pH was great and electron transfer from $O_2^{(\pm)}$ to the intermediate dioxetan (1) was thought to occur at high pH.

The main path is suppressed by prior nucleophilic attack by the t-butyl hydroperoxide.

This oxidation of adducts of lucigenin with nucleophiles is a common, if a rather complex, phenomenon. The first study was made by Janzen [7,8] of the reaction of lucigenin pseudo-base (4):

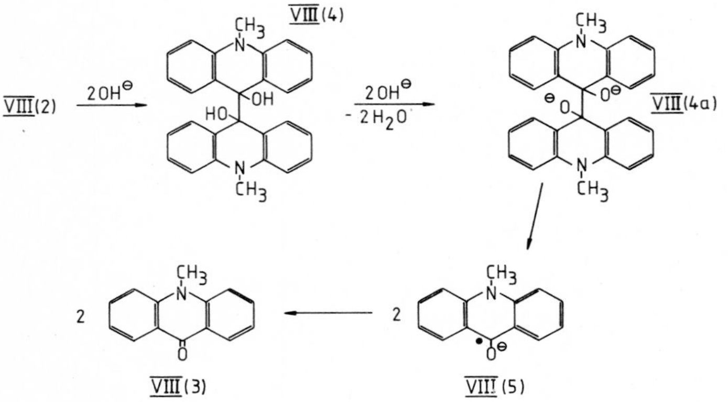

Although no ESR signal indicative of the N-methyl acridone radical anion (5) was observed, homolytic cleavage as shown was nevertheless assumed as being likely.

Signals diagnostic of the 10,10'-dimethyl-9-hydroxy-9,9' biacridan radical were seen. This radical may arise by electron transfer from (5) to lucigenin monocarbinol or lucigenin itself, followed by addition of hydroxyl ion. The actual excitation step is thought to involve electron transfer from (5).

The weak chemiluminescence from lucigenin in alkaline DMSO solutions in the absence of O_2 may involve similar reactions. Cyanide ions also react with lucigenin to give light in the presence of oxygen [7]. Again N-methyl acridone is the emitter and the biacridanyl derivative (6) appears to be an intermediate. An equilibrium between (6) and the radical (7) is proposed. Addition of cyanide ion to N-methyl acridinium chloride in air-saturated DMSO or DMF produces the same radical.

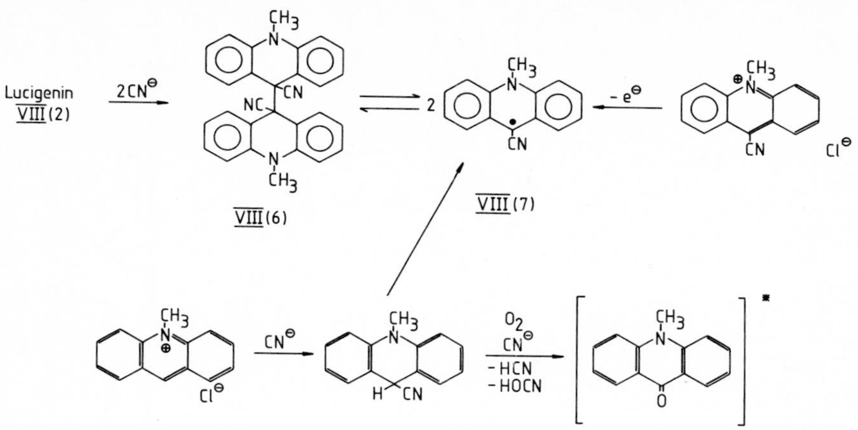

The presence of various reduced products such as 10,10'-dimethyl-9,9'-biacridylidene supports such a sequence with radical cations as intermediates.

Nucleophiles other than hydroxyl ion or cyanide such as trifluoroethoxide or ethylamine give similar results. The radical cation (8) is a significant intermediate, capable of reacting with either O_2 or $O_2^{(\div)}$.

Several lucigenin byproducts were identified, and ESR (at pH 13) demonstrated the presence of radicals formed independently of the presence of oxygen. Such signals disappeared rapidly on the addition of H_2O_2. Reaction of O_2 with the lucigenin radical cation (i. e. lucigenin plus one electron) apparently gave N-methyl acridone.

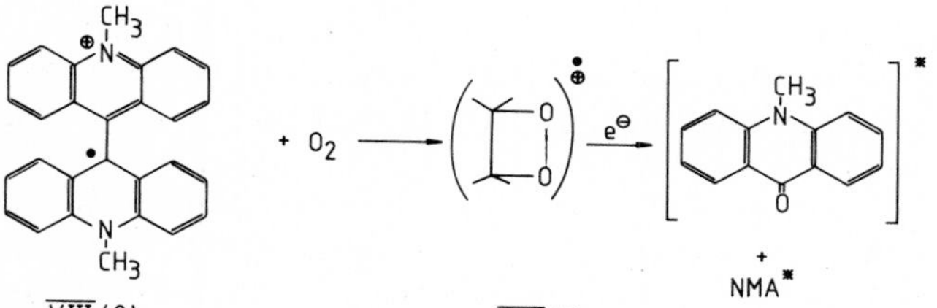

Reaction with superoxide ion is no doubt responsible for the observed chemiluminescence of lucigenin in the presence of xanthine and xanthine oxidase [10, 11].

VIII.1.2 Energy Transfer in Lucigenin Chemiluminescence

Finally, the existence of the potentially confusing energy transfer* has been demonstrated by showing that the green emission is dependent on the concentra-

---

* As early as 1943 Kautsky [18] provided experimental evidence for such an energy transfer in the lucigenin chemiluminescence.

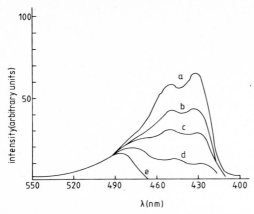

Fig. 14. Effect of lucigenin (L) solutions of varying concentrations upon the emission spectrum of N-methyl acridone (3). After Maskiewicz et al. [9]. Concentrations of L: *a)* 0.0 M; *b)* 2×10^{-5}M; *c)* 4×10^{-5}M; *d)* 6.6×10^{-5}M; *e)* 2×10^{-4}.

tion of lucigenin (Fig. 14). Kinetic analysis is in accord with the mechanism on p. 111.

VIII.1.3 Influence of Micelle Forming Agents on the Lucigenin Reaction

A distinct increase in the chemiluminescence quantum yield of the lucigenin reaction was observed in the presence of cetyltrimethylammonium bromide (CTAB), compared with the reaction performed in an isotropic medium [12]. The effect of CTAB depends on the ratio of the lucigenin/CTAB-concentration.

The maximum effect (which gives ca. 4 times the quantum yield of the chemiluminescence of isotropic solutions) is reached when the ratio is 0.2, with a CTAB concentration of 1×10^{-1} mol.

It is assumed that the primary excited N-methyl acridone is solubilized in the micelle. Thus energy transfer to other acridine derivatives, present in the alkaline solution, is inhibited. The N-methyl acridone spectrum is visible in the lucigenin chemiluminescence emission when CTAB is present, relatively unaffected by the more usual accompanying green emission.

VIII.2 Acridinium Salts and Acridans

The acridine nucleus has proved to be a good basis for a variety of chemiluminescent compounds. Among these are the acridinium active esters which not only provide the best understood examples of chemiluminescence to date but are of considerable value in determining the mechanism of light emission in certain bioluminescent organisms.

Acridinium phenyl esters have been thoroughly investigated [14, 19, 22] with the convincing mechanism shown below as a result:

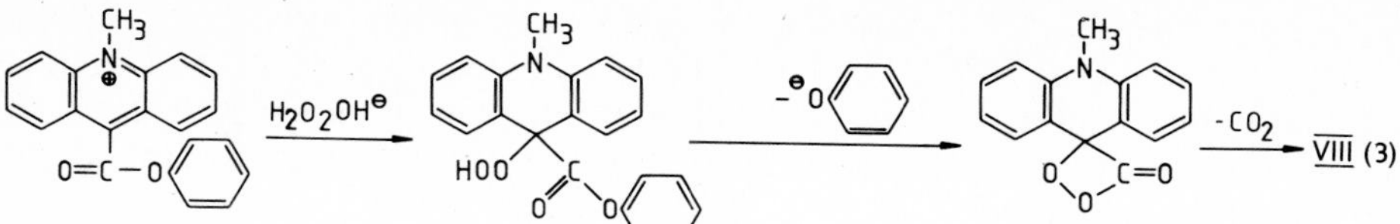

The scheme at the top shows structures VIII(9), an intermediate with HOO group, VIII(13), and leading to VIII(3) with loss of CO_2. Reagents include $H_2O_2OH^{\ominus}$ and $^{\ominus}O$-phenyl.

The first equilibrium is very pH sensitive, but hydrogen peroxide is a most effective competitor for hydroxyl ion, even when present in very low concentrations. Any leaving group with a pK_a for the conjugate acid below about 11 is an effective substituent in chemiluminescence. As well as phenols (particularly those with electron withdrawing substituents), fluoroalcohols and thiols give efficient derivatives. Hydrolysis is a major source of loss of efficiency, since the carboxylic acid (reacting as shown) is essentially non-chemiluminescent. Alkyl hydroperoxides, incapable of forming the four-membered ring, do not yield chemiluminescence. Aliphatic esters suffer hydrolysis rather than attack by the peroxide. The excitation quantum yield is about 5%.

A similar mechanism applies to the acridine phenyl esters and acridine N-oxide phenyl esters [21]. Quantum yields are not quite so high, and in the former case, dipolar aprotic solvents are required, presumably to enhance nucleophilic attack at the now less activated 9-position. Quinolinium (10) and phenanthridinium (11) esters are also chemiluminescent but with a lower efficiency [21]:

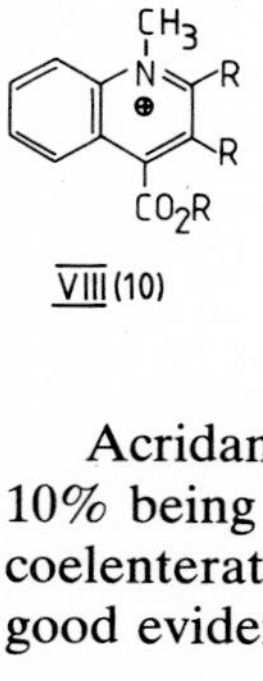

Structure VIII(10): a quinolinium with CH3 on N, R substituents and CO_2R.

Structure VIII(11): a phenanthridinium with CH3 on N and CO_2Ar.

Acridan active esters are particularly efficient, quantum yields in the region of 10% being obtainable. The reaction is entirely analogous to that of firefly and coelenterate luciferins. Dipolar aprotic solvents give the best results and there is good evidence for all the steps shown [21]:

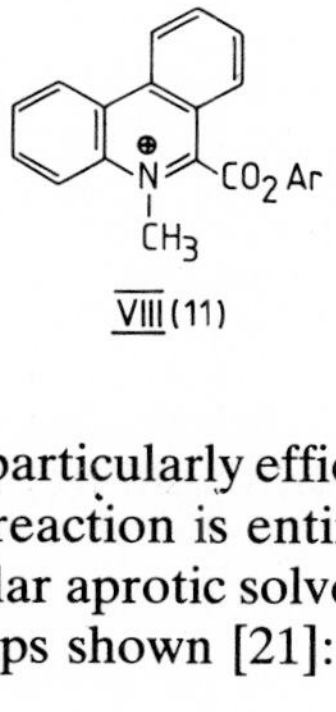

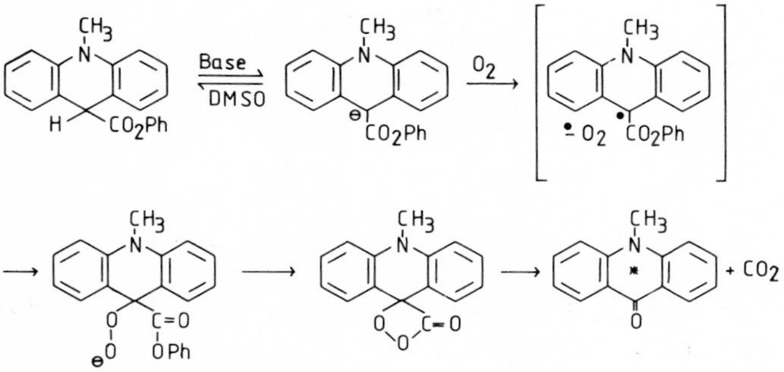

Reaction scheme showing acridan with H and CO_2Ph reacting with Base/DMSO, then O_2, through intermediates $\cdot O_2$ CO_2Ph, to a dioxetanone, and finally to excited acridone $+ CO_2$.

These compounds all react by formation of a dioxetanone. This is the immediate precursor of the light emitting step. It is significant that the peroxide (12) is isolable, yet the dioxetanone is certainly not. The formation of the strained per-ester increases the oxidizing potential of the peroxide to the point where electron transfer from the electron rich heterocycle ensues:

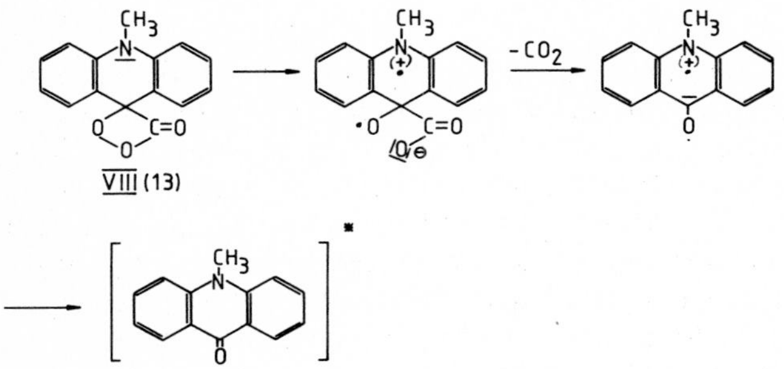

VIII (13)

This intramolecular version of CIEEL has been invoked for the luciferins [16 b] and electron rich dioxetans [16 a].

The acridine nucleus has been used in the examination of hydrazide chemiluminescence as previously discussed, but several other substituents can result in chemiluminescent compounds. Although the aldehyde (14, R = H) can react via the anion (17) the ketone (15) cannot [15].

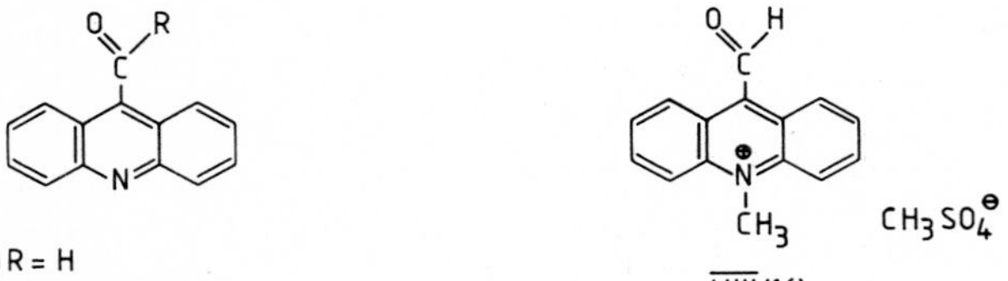

VIII (14) R = H

VIII (15) R =

VIII (16)

An alternative mechanism is shown, and has been encountered previously (p. 105).

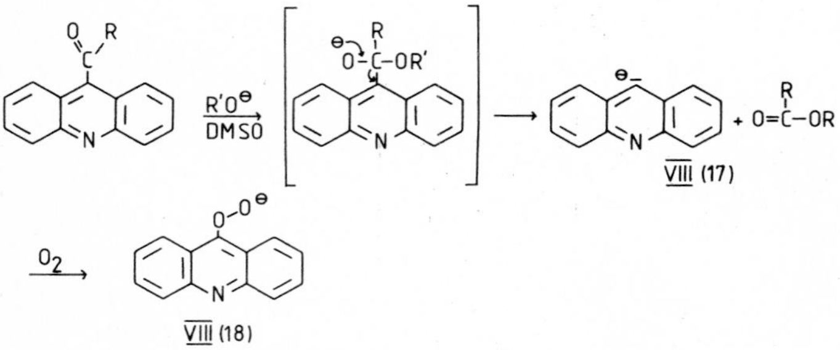

VIII (17)

VIII (18)

A similar mechanism produces an alternative to the acyl anion or dioxetan reactions for the aldehyde (16).

A very high chemiluminescence quantum yield (about eight times that of luminol) has been reported for the oxidation of N-methyl-9-benzyl acridinium chloride with potassium persulfate as catalyst [17]. Although no mechanism has been determined the reaction is reminiscent of the oxidation of 9-methyl acridine [15] in which the carbanion reacts with oxygen to produce the dioxetan which presumably decomposes in the accepted fashion for such compounds:

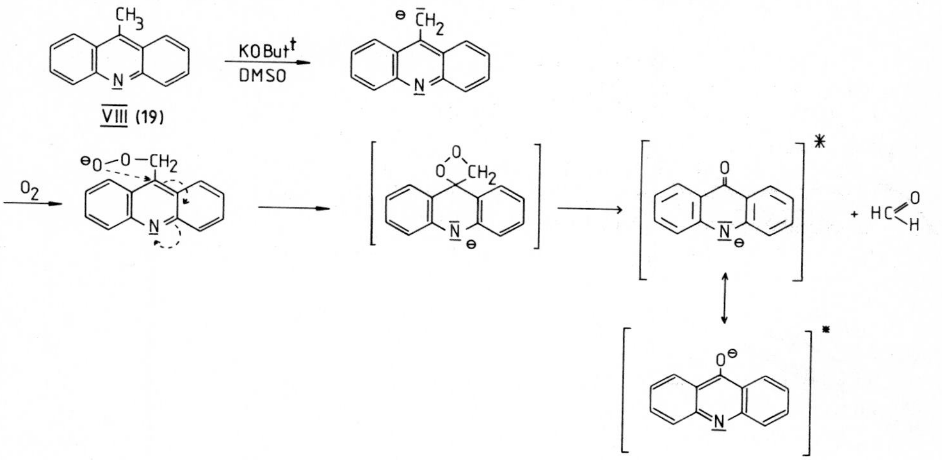

It is very difficult to follow these alternatives to the very well defined dioxetan route right through to the light emitting step. A variety of fates for the intermediate peroxide, and a variety of electron transfer mechanisms are possible. Very many heterocycles and fluorescent aromatic hydrocarbons are chemiluminescent in dipolar aprotic solvents with base in the presence of oxygen. The quantum yields are rarely high, and the identification of a single well defined pathway is extremely difficult.

VIII.3 References

1. Gleu, K. and Petsch, W., Angew. Chem. 48, 57 (1935)
2. Gundermann, K.-D., Chemilumineszenz organischer Verbindungen p. 90, Springer-Verlag, Berlin 1968
3. McCapra, F. and Richardson, D. G., Tetrahedron Lett. 1964, 3167
4. McCapra, F. and Hann, R. A., JCS, Chem. Commun. 1969, 442
5. Wang, K. W., Singer, K. D. and Legg, K. D., J. Org. Chem. 41, 2685 (1976)
6. Janzen, F. G., Pickett, J. B., Happ, J. W. and de Angelis, W., ibid., 35, 88 (1970)
7. Happ, J. W. and Janzen F. G., ibid., 35, 96 (1970)
8. Happ, J. W., Janzen, F. G. and Rudy, B. C., ibid. 3382 (1970)
9. Maskiewicz, R., Sogah, D. and Bruice, T. C., J. Amer. Chem. Soc. 101, 5347, 5355 (1979)
10. Legg, K. D. and Hercules, D. M., ibid., 91, 1902 (1969)
11. Greenlee, L., Fridovich, I. and Handler, P., Biochemistry 1, 779 (1962)

12. Paleos, C. M., Vassilopoulos, G. and Nikokavouras, J., J. Photochemistry *18,* 327 (1982)
13. Rauhut, M. M., Sheehan, D., Clarke, R. A., Roberts, B. G. and Semsel, A. M., J. Org. Chem. *30,* 3587 (1965)
14. McCapra, F., Richardson, D. G. and Chang, Y. C., Photochem. Photobiol. *4,* 1111 (1965)
15. Rapaport, E., Cass, M. W. and White, E. H., J. Amer. Chem. Soc. *94,* 3160 (1972)
16 a. Koo, J.-Y., Schmidt, S. P. and Schuster G. B., Proc. Nat. Acad. Sci USA *15,* 30 (1978)
16 b. McCapra, F., Beheshti, I., Burford, A., Hann, R. A. and Zaklika, K. A., JCS Chem. Commun. *1977,* 944
17. Gaglias, J. and Nikokavouras, J., Monatsh. Chemie *110,* 763 (1979)
18. Kautsky, H. and Kaiser, H., Naturwiss. *31,* 505 (1943)
19. McCapra, F. and Chang, Y. C., JCS Chem. Commun. 1966, 522
20. McCapra, F., Richardson, D. G. and Hann, R. A., to be published.
21. McCapra, F., Progr. Org. Chem. *8,* 231 (1973)
22. McCapra, F., Taheri, M. and Perring, K. D., unpublished results.

IX. Other Nitrogen-Containing Compounds – Imine Peroxides

Among the peroxides in general there are several classes of compounds which contain the structural element A derived from compounds such as tetrakis-dimethylamino ethylene (TMAE), Schiff bases, diisoquinolinium salts, indoles etc. Although the mechanisms are probably related, there are several areas of doubt in interpretation. Many of the observations are consistent with the intermediacy of a dioxetan, but in view of the expected short lifetimes in the presence of electron donating nitrogen, confirmation is difficult. Indeed while such electron donation and electron transfer can be involved to explain such short lifetimes, the explanation is not complete.

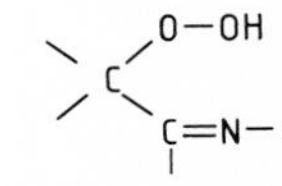

IX.1 Tetrakis-dialkylaminoethylenes

The first member of this series of compounds, tetrakis-dimethylaminoethylene (TMAE) (1) is commercially available, and is an unique example of a simple compound spontaneously chemiluminescent on exposure to oxygen or air [1,2]. This property has led to a large number of patent applications for uses such as a lighting source, fish lures or warning markers.

The major product is the tetramethylurea (2), and the most likely route to its formation is shown:

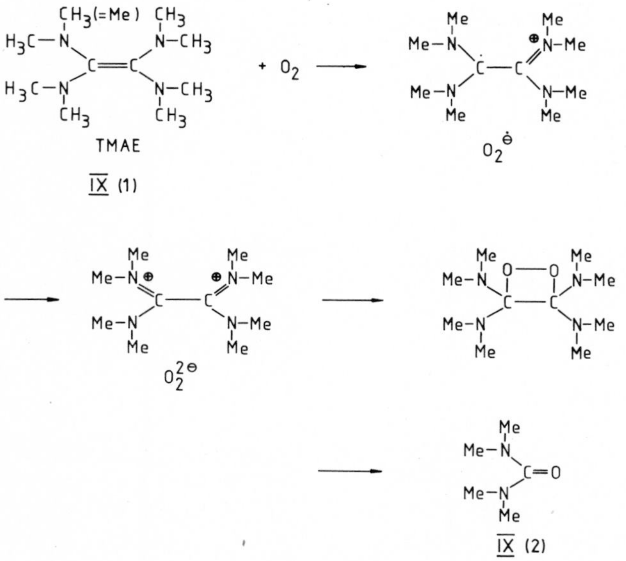

The emitter is TMAE itself, and this fact obscures the primary excited state [1]. It is likely that TMAE is excited by energy transfer.

No effect of added fluorescers is observable, almost certainly because TMAE has a very low ionisation potential and will quench all excited states. In addition, the reaction occurs best in neat TMAE, enhancing the quenching effect.

An alternative explanation [3] which recognises the apparent second order kinetics of light emission is shown below:

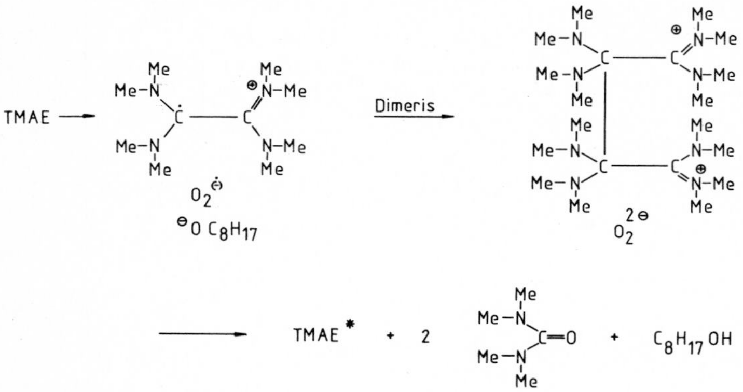

The reaction is very sensitive to solvent, and hydroxylic solvents (but not water) are catalytic for the reaction. It is not clear how the population of the excited state is achieved. It must be noted that there are several intermediates and by-products detected by ESR and NMR. The light emitted (from TMAE) is green (λ_{max} 500 nm) [1, 2] with a variable quantum yield (3.0×10^{-4} to 2.1×10^{-3} einstein M^{-1}).

The urea (2) ist a potent quencher of the emission, and some patents [4] describe means of absorbing it as it is produced.

Other alkyl tetra-amino-ethylenes are luminescent (e. g. (3) R = alkyl) but aryl substitution (3, R = Ph) results in a virtually non-luminescent compound. [2]

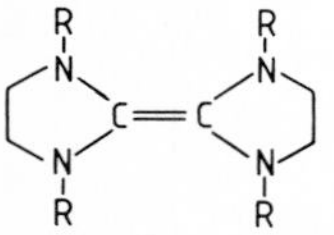

The explanation for this is not clear, although the electronegative phenyl group may affect the oxidation or catalyse intersystem crossing in the excited state.

IX.2 Bis-Isoquinolinium Salts

The chemiluminescence of these compounds was discovered [6] during an investigation into hindered rotation of the biphenyl sort. The parent compound [4] (n = 2) has a high quantum yield [10] (11×10^{-2} einstein M^{-1}), and the spectrum matches the fluorescence of 1,2-bis (isoquinolono) ethane (5) (n = 2)

IX (4) IX (5)

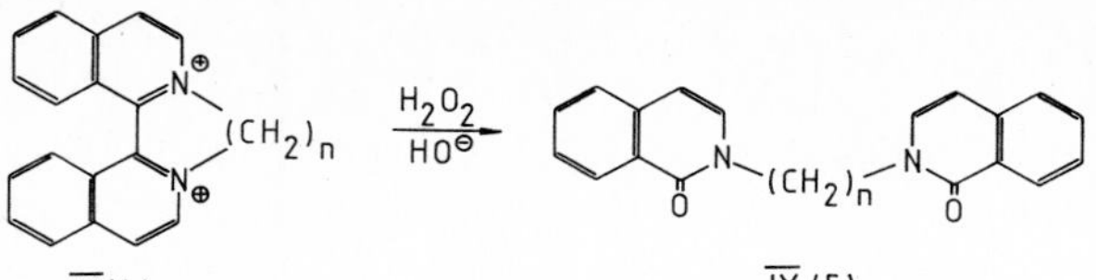

Other values of n also give chemiluminescent compounds, with an interesting but unexplained variation in quantum yields and spectra. The reaction mechanism has been studied [7, 8] with evidence provided for the suggestion [6] that the reaction occurs via the electron rich olefin (6) and presumably via the dioxetan (7).

IX (6)

IX (7)

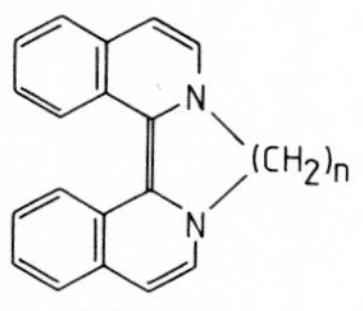

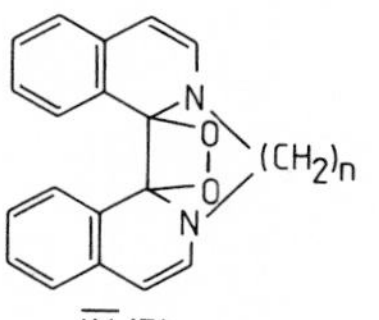

IX.3 Lophine and Related Compounds

This was the first discrete chemiluminescent organic compound, described in 1877 [9]. It was re-examined as the peroxide (9), by two groups [10, 11].

$R^1 = R^2 = \text{Phe}$

IX (8) IX (9)

IX (9a) R^1 = R^2 = Phe

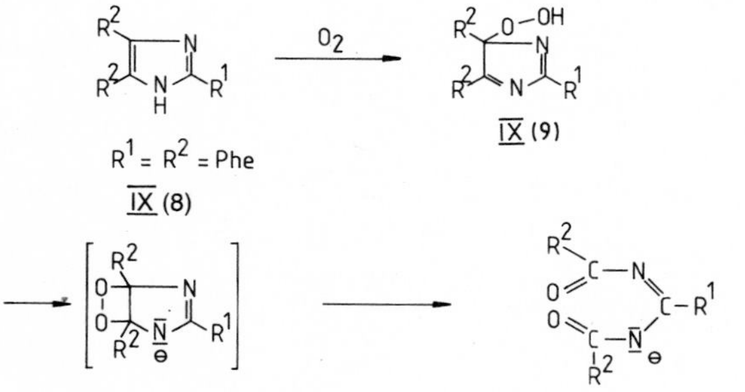

A fairly bright emission is observed at about 500 nm, but interestingly the major product (the dibenzoylbenzamidine) is not fluorescent under any conditions. A suggestion [11] has been made that the benzamidine is formed in a conformation related to the dioxetan, and that only this transient structure is fluorescent. However, a good match between the fluorescence of the dicarbonyl product is obtained for (9 a) with λ_{max} 490 nm, and a quantum yield of 10^{-3} to 10^{-4} einstein M^{-1} for this and similarly substituted compounds [11].

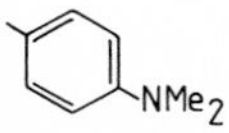

IX.4 Indole Derivatives

The chemiluminescent reactions of indoles were originally investigated in the mistaken belief that the tryptophan residue of *Cypridina* luciferin was the site of the light emitting reaction [12]. A large number of indoles were surveyed for luminescence in DMSO/t-BuOK in air but no mechanistic conclusion was drawn. The reactions were only moderately efficient, (10) for example having a quantum yield of about 10^{-5}.

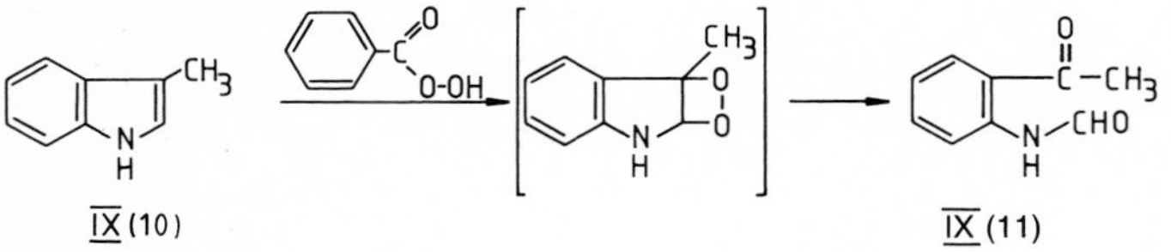

$\overline{\text{IX}}$ (10) $\overline{\text{IX}}$ (11)

The intermediate peroxide involved in the reaction was postulated and synthesised in another series of indoles (12) more suited to the isolation of the pure compounds [13]. The chemiluminescence of these peroxides, although not very efficient, is bright because of the fast reaction. This work established for the first time the unequivocal presence of a dioxetan on the reaction path by the use of $^{18}O_2$ and $H_2^{18}O$.

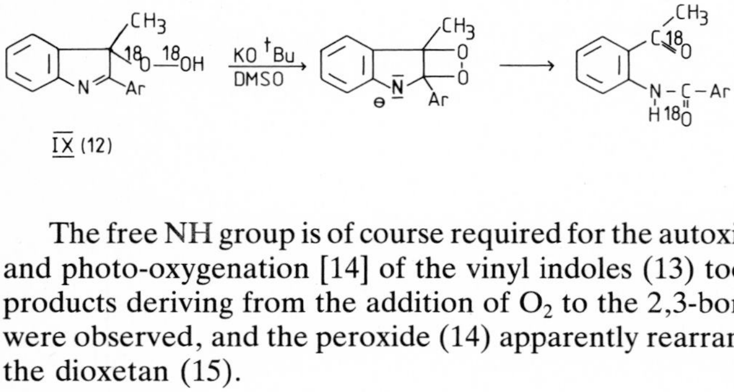

$\overline{\text{IX}}$ (12)

The free NH group is of course required for the autoxidation described above, and photo-oxygenation [14] of the vinyl indoles (13) took the route shown. No products deriving from the addition of O_2 to the 2,3-bond of the indole nucleus were observed, and the peroxide (14) apparently rearranges in protic solvents to the dioxetan (15).

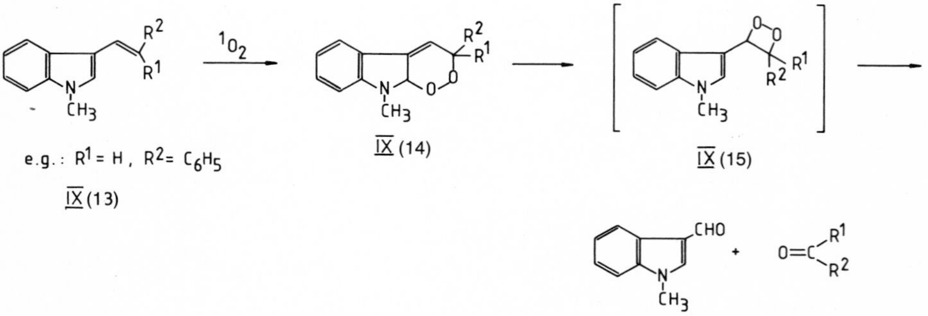

e.g.: $R^1 = H$, $R^2 = C_6H_5$

$\overline{\text{IX}}$ (13) $\overline{\text{IX}}$ (14) $\overline{\text{IX}}$ (15)

Oxidation of (10) by peracids in non-basic solution also results in chemiluminescence, and although a dioxetan is postulated, eventually yielding (11) only circumstantial evidence is available in this case [15].

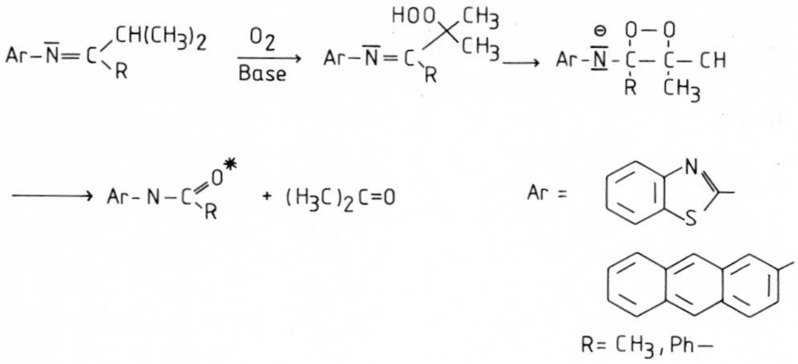

$$Ar = $$

R = CH_3, Ph—

IX.5 Chemiluminescent Schiff Bases

The simplest structures capable of forming imine peroxides are the Schiff bases. The earliest examples [16] were not at all efficient, but further development of the principle led to several very bright reactions. Some of the structures involved are shown below and a particularly noteworthy case is the trioxolan (16). This compound was at first thought [17] to be the dioxetan itself, but re-investigation [17, 18] resulted in the correct structure. One of these investigations [17] also demonstrated, by the use of $^{18}O_2$, that the isolated peroxide (16) first rearranged to a dioxetan in DMSO and strong base. The resulting decomposition provides a remarkably high quantum yield of 18%.

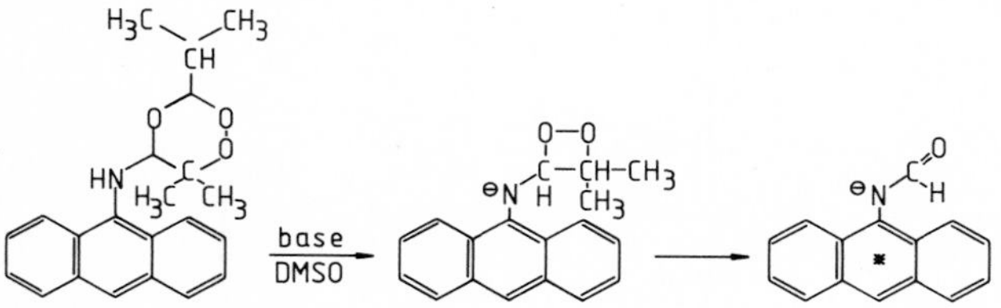

IX (16)

IX.6 References

1. Fletcher, A. N. and Heller, C. A., J. phys. Chem. 71, 1507 (1967)
2. Winberg, H. F., Carnahan, J. F., Coffman, D. D. and Brown, M., J. Amer. Chem. Soc. 87, 2055 (1965)
3. Urry, W. H. and Sheeto, J., Photochem. Photobiol. 4, 1067 (1965)
4. e. g. US. 3.888.785 (1970)
5. Kuwata, K. and Geske, D. H., J. Amer. Chem. Soc. 86, 2101 (1964)
6. Mason, S. F. and Roberts, D. R., JCS Chem. Commun. 1967, 476

IX. Other Nitrogen-Containing Compounds

7. Henry, R. A. and Heller, C. A., J. of Luminescence *4,* 105 (1971)
8. Heller, C. A., Henry, R. A. and Fritsch, J. M., in: Cormier, M. J., Hercules, D. M.and Lee, J. (Eds.) "Chemiluminescence and Bioluminescence" p. 249, Plenum Press, New York 1973
9. Radziszewski, B., Ber. dt. chem. Ges. *10,* 321 (1877)
10. White, D. M. and Sonnenberg, J., J. Amer. Chem. Soc. *86,* 5685 (1964)
11. White, E. H. and Harding, M. J. C., Photochem. Photobiol. *4,* 1129 (1965)
12. Cormier, M. J. and Eckroade, C. G., Biochim. Biophys. Acta *64,* 340 (1962); Philbrook, G. E., Ayers, J. B., Garst, J. F. and Totter, J. R., Photochem. Photobiol. *4,* 869 (1965)
13. McCapra, F., Richardson, D. G. and Chang, Y. C., Photochem. Photobiol. *4,* 1111 (1965); McCapra, F. and Chang, Y. C., JCS Chem. Commun. *1966,* 522
14. Matsumoto, M. and Kondo, K., J. Amer. Chem. Soc. *99,* 2393 (1977)
15. Omote, Y., Yamamoto, H., Funasaki, K., Akutagawa, M. and Sugiyama, Bull. Chem. Soc. Japan *42,* 3014 (1969)
16. McCapra, F. and Wrigglesworth, R., JCS Chem. Commun. *1969,* 91
17. Akutagawa, M., Aoyama, H., Omote, Y. and Yamamoto, H., JCS Chem. Commun. 1976, 180
18. McCapra, F., Chang, Y. C. and Burford, A., ibid., 1976, 608
19. Goto, T. and Nakamura, H., Tetrahedron Letts., 1976, 4627 Yamamoto, H., Aoyama, H., Omote, Y., Akuagawa, M., Takenaka, A. and Sasada, Y.; JCS Chem. Commun., 1977, 63

X. Miscellaneous Compounds

A very large number of organic compounds are weakly chemiluminescent, and if a possible mechanism is not apparent by inspection, they are difficult to classify. In this section are gathered a few distinctive compounds. Some of the mechanisms proposed will be familiar, whereas others are less well established.

X.1 Chlorinated Cyclic Carbonates, Glycolides, and Ethers

Tetrachloro ethylidene carbonate (1), on treatment with hydrogen peroxide and base in the presence of a fluorescer, yields chemiluminescence with a maximum quantum yield of ca. 6.5×10^{-2} einstein mole^{-1}, with emission from the fluorescer [1]:

$\underline{X}$ (1) $\underline{X}$ (2) $\underline{X}$ (3)

This chemiluminescent reaction was suggested as occuring via ring opening of the cyclic carbonate group with subsequent ring closure to the peroxide (2). An alternative is formation of the diperoxy acid (3). Although a relationship to oxalyl chloride and peroxyoxalate chemiluminescence seems reasonable, details of the excitation step are not known.

The perchloro-dioxolane (4), and the perchloro-1,4-benzodioxane (5) also produced light on reaction with H_2O_2 and fluorescer, but with lower quantum yields.

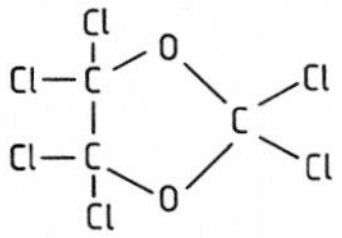

$\underline{X}$ (4) 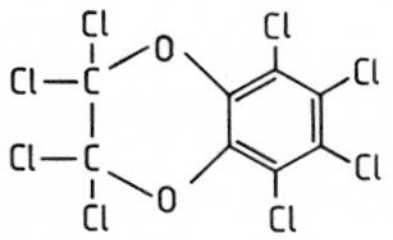 $\underline{X}$ (5) 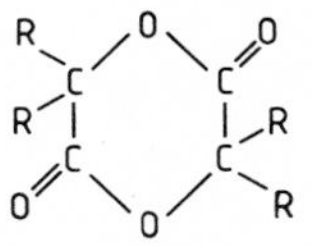 $\underline{X}$ (6)

R = Br, Cl, F
or
2R = alkyl, aryl, H

Similarly the chlorinated glycolides (6) have been reported to chemiluminesce under the conditions mentioned before [2].

125

X.2 Sulfuranes

The reaction of Martin's sulfurane [3], (7) which is used for the smooth dehydration of 1,2-diols to epoxides, with cumylhydroperoxide (8), hydrogen peroxide, or other hydroperoxides is chemiluminescent in the presence of DBA, or rubrene [4]:

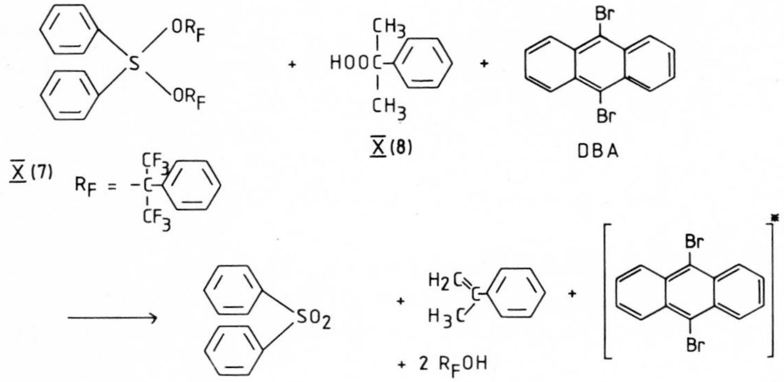

Two distinctly different stages could be observed in this new type of chemiluminescence: the first stage is quenched by oxygen, the second requires oxygen. NMR studies revealed that radicals occured as intermediates in these reactions.

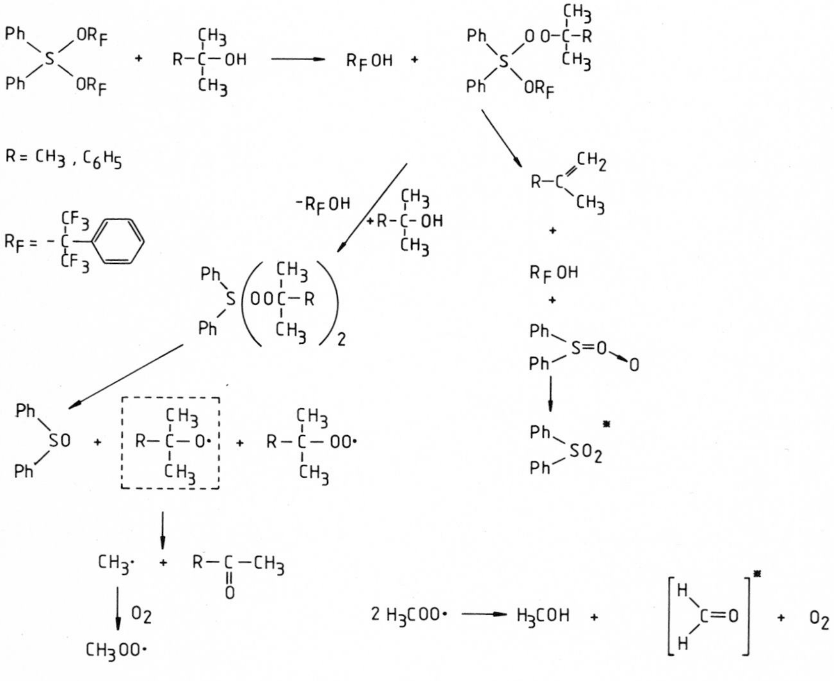

Excited diphenyl sulfone and excited formaldehyde might transfer their excitation energy to the fluorescer.

X.3 Thioesters

Apart from the tetrakis dialkylamino-ethylenes, the only other compounds which have been reported to be spontaneously chemiluminescent in air, are certain thioesters [5]. In contrast, the tetrakis alkylthio-ethylenes are not luminescent [6]. Sulphuric acid is the only product to be identified in what is evidently a gas phase reaction. The light intensity is roughly in proportion to the volatility of the esters (9–11). Although the report is sufficiently detailed as to describe the formation of luminescent smoke rings, attempts to reproduce the phenomenon have not been successful [7].

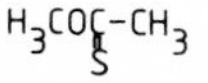

$H_3COC-CH_3$ $(CH_3)_2N-COCH_3$ $H_3CSC-OCH_3$

X (9) 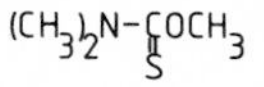   X (10) 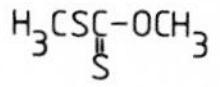   X (11)

X.4 Vinyl Halides

A very interesting chemiluminescent reaction not involving oxygen has recently been reported [8] in which Birch reduction gave rise to a luminescent inter-mediate.

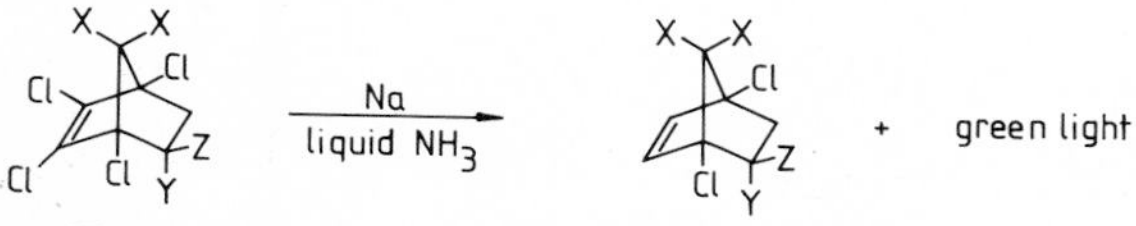

X (12)   X (13)

Light emission is completely dependent upon the presence of the vinylic chlorine substituents and on the fact that one of y or z must be an oxy-substituent. A variety of other relatively minor effects of substituents are noted.

The mechanism proposed is that the intermediate radical anion formed by the addition of an electron to the Π^* orbital of the double bond dissociated to give vinyl radical in its excited 2Π state and a chloride ion. The resulting 2Π to 2Σ (ground state) transition releasing 2.4 eV is consistent with the energy of the green emission. This is the result of a combination of electronic and geometric effects (e. g. the appearance of pyramidal character in the olefinic C atom) which favour a non-oxidative pathway.

X.5 Strained Hydrocarbon Isomerisation

A recently described [9] example of this chemiluminescence type is the rearrangement of lepidopterin (14) after irradiation:

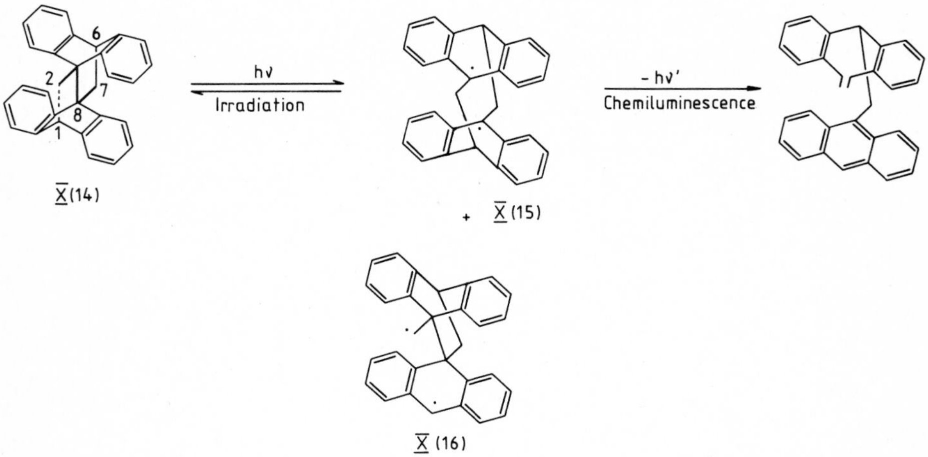

Interestingly a remarkably long-wave chemiluminescence emission (600–685 nm) was observed [9].

X.6 Metal Phosphides

The spontaneous chemiluminescence of white phosphorus in air is well known, and often considered as the archetype of chemiluminescence ("phosphorescence"). The reaction actually takes place in the gas phase close to the surface of the element. Among the emitting species are PO, $(PO)_2$ and HPO.

The mechanism is not fully understood, and it is not clear whether it can assist in explaining the more recently observed chemiluminescence of lithium alkyl phosphides [10].

Several examples were synthesized, lithium diphenyl phospide being among the brightest. It appears that a highly associated or perhaps even a solid phase is required for the luminescence. It does not appear to occur in solution and the fluorescence is that of the solid. The initial excited state has not been rigorously identified, although a mechanism is suggested:

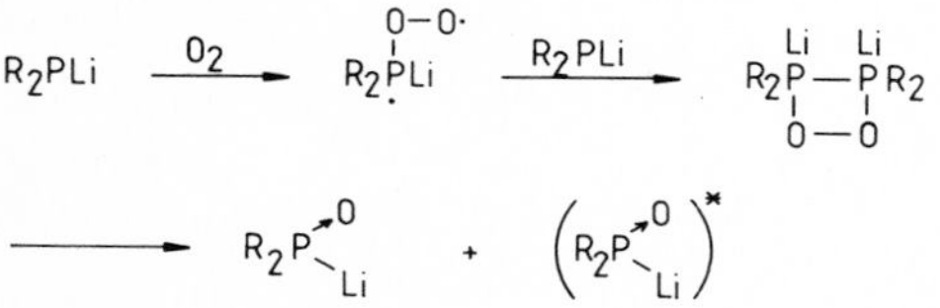

The first report on the chemiluminescence of lithium-dialkylphosphides was published by Issleib and Tzschach [11] in 1959.

X.7 References

1. Maulding, D. R. and Roberts, B. G., J. Org. Chemistry *37,* 1458 (1972)
2. US 3.697.432 (1969); American Cyanamid
3. Martin, J. C., Frary, J. A. and Arhart, J. R., J. Amer. Chem. Soc. *96,* 4606 (1974)
4. Bartlett, P. D., Aida, T., Chu, H.-K. and Fang, T.-S. ibid., *102,* 3515 (1980).
5. Delépine, M., Compt. rend. *154,* 1171 (1912)
6. Seebach, D., personal communication
7. McCapra, F., unpublished observations. White, E. H., personal communication
8. Cotsaris, E. and Paddon-Row, M. N., JCS Chem. Commun. *1984,* 95
9. Kaupp, K. and Schmitt, D., Chem. Ber. *113,* 3932 (1980)
10. Strecker, R. A., Snead, J. L. and Sollott, G. P., J. Amer. Chem. Soc. *95,* 210 (1973)
11. Issleib, K. and Tzschach, A., Chem. Ber. *92,* 1118 (1959); see also Issleib, K. and Döll, G., Chem. Ber. *94,* 2664 (1961); Issleib, K., and Krech, F., Chem. Ber. *94,* 2656 (1961)

XI. Electron Transfer Chemiluminescence

Although there are many components in a mechanistic description of a chemiluminescent reaction, the heart of the matter is the actual excitation step. Several such steps have been identified. Some are molecular in character e. g. the decomposition of dioxetans and some are intermolecular electron transfer steps. There is an intermediate class in which the step can be formulated as an *intra*molecular electron transfer. Many luminescent reactions have been ascribed to this category with varying degrees of confidence. Cyclic hydrazides such as luminol belong rather uncertainly here. Electron rich dioxetans and dioxetanones and the luciferins with such intermediates on the pathway are a little more reasonably assigned to an intramolecular electron transfer mechanism. Even here however caution is required in that *direct* evidence for *discrete* electron transfer will by its very nature be almost impossible to obtain and will probably remain circumstantial.

However it is possible to demonstrate excitation mechanisms which are unequivocally electron transfer in character. Both chemical and electrochemical methods can be used, and the observations in the two types can be nicely related. Extrapolation of these well founded results to intermolecular electron transfer where the reactants are normal electron paired species is also possible and has proved effective in codifying and indeed discovering a large range of CIEEL.

In this chapter, we shall consider different types of chemiluminescent radical ion reactions:
- annihilation reactions between radical anions and radical cations derived from the same parent compound, (e. g. DPA radical anion and DPA radical cation), or from different parent compounds (e. g. DPA and thianthrene)
- chemiluminescent reactions between a radical anion or a radical cation and an electron accepting or electron donating compound.

The simplest example of this type of radical ion chemiluminescence is the reaction of appropriate electron acceptors with hydrated electrons.

Experimentally, radical ions can be produced electrochemically – the chemiluminescence occurring called electrogenerated chemiluminescence (ECL) – or non-electrochemically by chemical reactions.

XI.1 Electrogenerated Chemiluminescence (ECL)

The first examples of this significant new form of chemiluminescence were reported in 1964 [1, 2, 3, 4]. Its importance lies in the precision with which the energy of the species involved can be measured and the simplicity of the electron transfer excitation step. In addition the generation of the reactive radical ions is easily controlled and followed by modern electrochemical techniques. In view of the central role that electron transfer reactions play in chemiluminescence generally, these advantages make the study of ECL attractive.

In its simplest and classical form a fluorescent aromatic hydrocarbon, such as 9,10-diphenyl anthracene (DPA) undergoes alternating oxidation and reduction at an electrode with an A. C. supply.

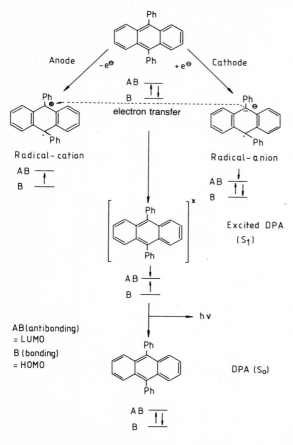

A sequence such as that in the scheme occurs. In principle this luminescence cycle can be repeated indefinitely, but in practice loss of DPA by side reactions, and oxidation of the solvent and supporting electrolyte interfere.

The formation of the radical anion/cation in the diffusion layer is followed by an annihilation reaction. If there is sufficient energy in this electron transfer to populate the excited state of the hydrocarbon, luminescence results. The energy available is determined by the redox potential of the ions.

XI.1.1 Energy-sufficient and Energy-deficient ECL

In all ECL reactions observed so far, the emission is a fluorescence, i. e. the first excited singlet state [51] of the emitter is involved.

The enthalpy H^O of the ECL redox process thus either must be sufficient to populate the S_1 state: "energy sufficient ECL", or, if it is not ("energy deficient ECL") there must exist an alternative mechanism to provide for the S_1-energy.

The standard enthalpy can be estimated from the cyclic voltammetric peak potentials (E_p) for systems unperturbed by decomposition or strong ion pairing of the reactants, and assuming an entropy term $T \Delta S$ of ca. 0.1 eV. Then we have:

$$- H^O = E^O (R^{(\dot{+})}/R) - E(R/R^{(\dot{-})}) - T \Delta S$$
$$= E^O (R^{(\dot{+})}/R) - E(R/R^{(\dot{-})}) - 0.16 \text{ eV}$$

The energy requirements are treated more extensively by Faulkner [3, 4] and Hoytink [6]. 9,10-Diphenylanthracene (DPA) is an example of an energy-sufficient system since for DPA, $E_p (R/R')^{(-)} = -1,89 \text{ V}$, $E_p (R/R^O)^{(+)} = + 1.35 \text{ V}$ and $E_s = 3.00 \text{ V}$.

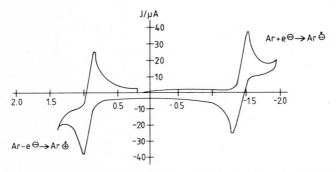

Fig. 15. Cyclic voltammetric curve. 0.593 mM rubrene in benzonitrile with 0.1 M TBAP.* Potential scan begins at + 0.2 V. (From [3]).

Cyclic voltammetry of the type depicted in Fig. 15 uses one electrode only which is made to act periodically as anode or as cathode. Although this is a most convenient technique it does have some disadvantages. The radical cation in particular is unstable and as the voltage changes during the other half-cycle, side reactions ensue [7]. Thus the ECL quantum yield in this single electrode technique is dependent on whether the radical cation or the radical anion is produced first.

A valuable new technique has been introduced by which both radical ion species can be produced simultaneously in a very small volume – by rotation of the electrodes, the diffusion of the radical ions towards each other is enhanced [8]. This is achieved by the special design seen in Fig. 16 (Rotating Ring Disc-Electrode; RRDE).

---

* TBAP: Tetrabutylammoniumperchlorate.

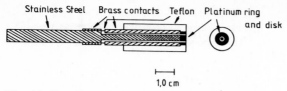

Fig. 16. Rotating-Ring-Disc-Electrode (after [8]).

The first use of this electrode (in non-luminescent electrochemistry) was reported in 1959 [9]. Its advantages are that as both radical ion species are produced at the same time there is no loss of that species produced during one half-cycle at a single electrode. Accumulation of side products in the vicinity of the electrode is prevented and steady light levels can be obtained. This is not possible at a single electrode.

Its use in ECL has been described [8]. Actinometric calibration of the RRDE has been achieved [10], and thus the efficiency of ECL can be effectively measured.

In light generation by cyclic voltammetry "pre-annihilation" chemilumines-cence ist observed, in which light is produced before the electrode reaches a potential sufficient to supply the second radical ion. This is avoided by using the RRDE in conjunction with carefully purified reagents and solvents.

ECL can also be obtained when the reactants in the redox reaction arise from different parent molecules (so-called mixed systems). The system N,N,N',N'-tetramethyl p-phenylene-diamine/DPA is an example. This is an energy-deficient system. Another very efficient mixed system is that of 9,10-dimethyl-anthracene radical anion (1) and tri-p-tolylaminium perchlorate (2).

$$\left(\begin{array}{c}\text{CH}_3\\ \\ \text{CH}_3\end{array}\right)^{(\pm)} \qquad \left(\text{H}_3\text{C}-\!\!\!\!\!\!\!\!\!\!\!\!\right)_3 \overset{\oplus}{\text{NH}} \quad \text{ClO}_4^{\ominus}$$

$$\underline{\text{XI}} \, (1) \qquad\qquad\qquad \underline{\text{XI}} \, (2)$$

Luminescence in this system was first investigated by a nonelectrolytic technique [3].

DPA-thianthrene (3) or 2,5-diphenyl oxadiazole (4)-thianthrene are energy-sufficient systems:

$$\underline{\text{XI}} \, (3) \qquad\qquad \text{H}_5\text{C}_6 \overset{\text{N}-\text{N}}{\underset{\text{O}}{\diagup \diagdown}} \text{C}_6\text{H}_5$$

$$\underline{\text{XI}} \, (4)$$

The emission spectrum matches the fluorescence of both thianthrene (3) and 2,5-diphenyl oxadiazole (4). The radical anion and cation of both compounds are involved.

XI.1.1.1 Energy-deficient ECL

Since the energy of singlet states required for chemiluminescence is so high, in many cases the redox potential of the reactants is insufficient to provide direct access to the S_1 state.

However even in these cases light emission is easily observed. The singlet-triplet splitting in most of the examples is large, and triplet states are usually accessible. Triplet-triplet annihilation then ensues to populate the singlet state as shown:

$$Ar^{(+)} + Ar^{(-)} \rightarrow {}^3Ar^* + Ar \quad \text{(Formation of excited triplet state)}$$
$${}^3Ar^* + {}^3Ar^* \rightarrow {}^1Ar^* + Ar \quad \text{(T-T- annihilation)}$$
$${}^1Ar^* \rightarrow Ar + h\nu$$

(Ar is used here for an aromatic hydrocarbon as well as for a fluorescent heterocyclic compound).

The abbreviation "T-route" has been used for this excitation pathway [14] in distinction to the direct excitation of a singlet state ("S-route") formed by

$$R^{(+)} + R^{(-)} \rightarrow {}^1R^* + R$$
$${}^1R^* \rightarrow R + h\nu$$

Excimer formation, as shown by the typical longwavelength structureless band of Fig. 17 (p. 135) is taken as evidence for the annihilation process. These excimers are observed even in polar solvents, where excimer formation by the normal fluorescence route is negligible. The formation of the singlet from two indentical precursors in a solvent cage is thus indicated. Excimers between non-identical partners is also observed.

$$Ar^{(+)} + Ar^{(-)} \rightarrow {}^1Ar_2^* \quad \text{(Excimer (= excited dimer) formation)}$$
$${}^1Ar_2^* \rightarrow 2\,Ar + h\nu \quad \text{(Excimer fluorescence)}$$
$$Ar^{(+)} + P^{(-)} \rightarrow {}^1(ArP)^* \quad \text{(Formation of a fluorescent exciplex = excited complex with a compound P)}$$

or:

$$Ar^{(-)} + P^{(+)} \rightarrow {}^3Ar^* + P \quad \text{(Formation of excited triplet states of the donor: T-T annihilation of } {}^3Ar^*)$$
$$Ar^{(+)} + P^{(-)} \rightarrow {}^3Ar^* + P \quad \text{(Formation of excited triplet states of the acceptor; subsequent T-T annihilation)}$$

Triplet formation has been proved by several independent experimental methods [3] among which are the influence of an external magnetic field and the effect of triplet quenchers on the chemiluminescence of an energy-dificient system.

There should be no observable effect of a magnetic field on energy-sufficient or "marginal"*) systems.

However, the ECL-intensity of energy-deficient systems is increased as paramagnetic quenchers, particularly molecular oxygen, are prevented from reacting with the triplet species leading to ECL by the T-route [3].

Triplet quenchers cause a decrease of ECL in energy-deficient systems, and trans-stilbene can be transformed into the cisisomer. This type of quenching is especially efficient when the energy of the triplet state of the primary excited product is ca. 0.2 eV higher than that of trans-stilbene [15].

---

*) These interesting systems have a redox enthalpy just below that required to populate the S_1 state directly. This threshold region is not yet fully understood [3].

When the quenchers are fluorescent themselves, e. g. anthracene and pyrene, the fluorescence of these compounds is emitted [16] as found in the fluoranthene/ 10-methyl-phenothiazine (5) system (Fig. 17)

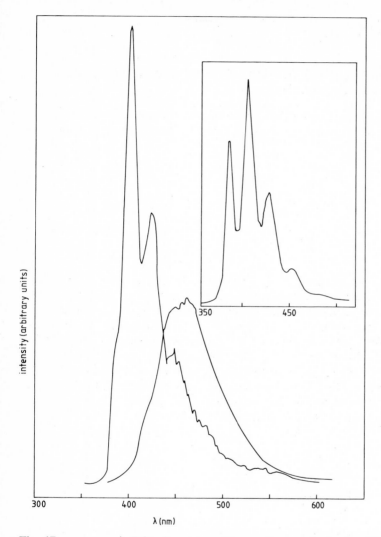

Fig. 17.

XI(5)

a) ECL- Spectrum of fluoranthene radical anion / 10-methylphenothia-zine in DMF. Concentrations: 0.001 M of each reactant. Alternating steps at $- 1.75$ V and $+ 0.88$ V vs. SCE.

b) ECL spectrum obtained under the same conditions, but in the presence of 0.001 M anthracene, the fluorescence spectrum of which is shown in the inset (10^{-5} M in DMF). Re-absorption reduces the O,O intensity in b) [16].

135

In the presence of anthracene (which has an energetically lower triplet state than fluoranthene or 10-methyl-phenothiazine), the ECL emission is shifted from the broad maximum at ca. 450 nm to the characteristic anthracene fluorescence spectrum with the three maxima at 390, 410 and 445 nm.

Triplet states can also be detected by excimer emission, and the study of luminescence transients [3].

XI.1.2 Chemiluminescent Reactions of Electrogenerated Radical Anions with Organic Halides.

DPA, rubrene, fluorene and other fluorescent aromatic hydrocarbons are reported to yield chemiluminescence when electrolyzed together with alkyl and aryl halides such as (6) or (7)

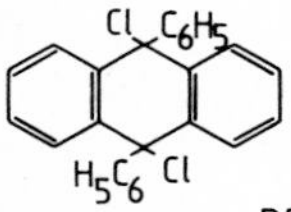

= DPACl$_2$

$\underline{XI}$ (7)

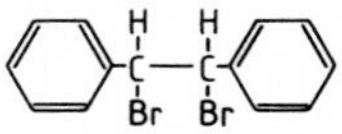

$\underline{XI}$ (6)

Electrolysis of mixtures of DPACl$_2$ (6) and rubrene produces rubrene fluorescence at the reduction potential of rubrene anion formation (-1.4 V); at -1.9 V, the reduction potential of DPA radical anion, mixed emission of rubrene and DPA fluorescence are observed. In all cases investigated so far, ECL of mixtures of aromatic hydrocarbons and halides like (7) shows an emission matching the fluorescence of that hydrocarbon which is most easily reduced [17].

A more detailed investigation of this ECL [30 A] led to the proposal of the following mechanism, which includes one-electron transfers from the hydrocarbon and the radical anion derived therefrom to the alkyl halide, so that, finally, the corresponding hydrocarbon radical cation is produced:

1) $R + e^{(-)}$ $\rightleftharpoons$ $R^{(\pm)}$
2) $R^{(\pm)} + AX_2$ $\rightleftharpoons$ $AX_2^{(\pm)} + R$
3) $AX_2^{(\pm)}$ $\rightarrow$ $AX\cdot + X^{(-)}$
4) $R^{(\pm)} + AX\cdot$ $\rightleftharpoons$ $1\ R^* + AX^{(-)}$
5) $R + AX\cdot$ $\rightleftharpoons$ $R^{(+)} + AX^{(-)}$
6) $AX^{(-)}$ $\rightarrow$ $A + X^{(-)}$
7) $R^{(+)} + R^{(\pm)}$ $\rightarrow$ $^1R^* + R$
8) $^1R^*$ $\rightarrow$ $R + h\nu$

(R = fluorescent aromatic hydrocarbon, AX$_2$ = alkyl halide)

XI.1.3 Efficiency of ECL Reactions

As in other types of chemiluminescence it is necessary to distinguish between chemiluminescence quantum yield ($\emptyset_{CL}$) and excitation yield ($\emptyset_{ES}$). In the measurement of $\emptyset_{CL}$ many experimental difficulties have to be overcome such as low intensities, the anisotropic nature of the light source and problems in accommodating the rather complex cell within the light detection chamber. Such measurements have been made possible by the RRDE technique. The highest quantum yields so far obtained (8.7×10^{-2} einsteins/mole) are for the rubrene radical ion annihilation in benzonitrile. Other ECL reactions are distinctly less efficient as table 11 demonstrates. An excitation yield ($\emptyset_{ES}$) of $1.5–11 \times 10^{-2}$ for the direct population of a singlet state by DPA radical ion annihilation has been recorded [13].

For triplet states, excitation yield values up to unity were measured for the energy-deficient system fluoranthene/10-methyl-phenothiazine [14, 18]. The estimation of the triplet yield was carried out by interception with cis-stilbene (p. 57).

XI.1.4 Ruthenium Chelate Complexes

The redox reactions of metals are classical electron transfer reactions. However few transition metals or their complexes are fluorescent in solution. A well known exception is that of ruthenium. Chemiluminescence was first reported in 1971 [19] from the chemical reduction and ECL demonstrated a short time later [20], with the reaction:

$$\text{Ru (bipy)}_3^{1(+)} + \text{Ru(bipy)}_3^{3(+)} \rightarrow \text{Ru (bipy)}_3^{2(+)*} + \text{Ru(bipy)}_3^{2(+)} \text{ (A)}$$

The behaviour of $\text{Ru(bipy)}_3\text{Cl}_2$ in cyclic voltammetry (acetonitrile solution with tetrabutyl- ammonium perchlorate as supporting electrolyte) was explained by several one-electron transfer steps in the potential range of $+1.70\,\text{V}$ and $-2,22\,\text{V}^*$). This corresponds to an oxidation to a $\text{Ru (bipy)}_3^{3(+)}$ species followed by stepwise reduction to the $+1,0$, and -1 charged species of Ru(bipy)_3 chelate complex between -1.09 and $-1.53\,\text{V}$ until, at $-2,22\,\text{V}$, free 2,2' bipyridine is liberated from the complex.

Between potentials of $+1,75\,\text{V}$ and $-1,60\,\text{V}$ of a cyclic square wave, an orange chemiluminescence (λ_{max} ca. 610 nm) is observed.

The enthalpy of the redox reaction in equation (A) is $2,6\,\text{eV}$ and is only marginal for direct excitation of the excited singlet state of $\text{Ru(bipy)}_3^{2(+)*}$, but it is energy-sufficient for the excited triplet state.

This first example of ruthenium chelate ECL was followed by investigations of several other ruthenium complexes, the ligands of which were (8) [21]:

bipy   XI (8)   XI (9)   XI (10)

---

* against an Ag-reference electrode.

Table. 10. Energy and Kinetic Relationships in ECL (Energy in V; after Hercules [1])

Compound	$(E_{1/2})_{Ox}$ Ar → Ar⁺	$(E_{1/2})_{Red}$ Ar → Ar⁻	ΔE*	ΔE_{corr}†	Energy of Lowest Excited Singlet State	Energy Deficit‡	$\emptyset_{max}$ at 25°C§	$k_{6(max)}$‖
Rubrene	+0.88	−1.47	2.35	2.15	2.30	−0.15	0.004	4×10^7
9,10-Diphenyl anthracene	+1.32	−1.72	3.04	2.84	2.85	−0.01	0.67	7×10^9
9,10-Dimethyl anthracene	+1.16	−1.82	2.98	2.78	3.05	−0.27	3×10^{-5}	3×10^5
Anthracene	+1.29	−1.95	3.24	3.04	3.29	−0.25	6×10^{-5}	6×10^5
1,3,4,7-Tetraphenyl isobenzofuran	+0.92	−1.96	2.88	2.68	2.51	+0.17	1.00	10^{10}
N-Methyl-1,3,4,7-tetraphenyl-isoindole	+0.66	−2.51	3.17	2.97	2.89	+0.08	1.00	10^{10}

* $\Delta E = (E_{1/2})_{Ox} - (E_{1/2})_{Red}$.

† $\Delta E_{corr} = \Delta E - T\Delta S$. In general, the $T\Delta S$ term has been estimated to be ca. 0.2 V.

‡ The energy deficit is the difference between ΔE_{corr} and the energy of the lowest excited singlet state.

§ This is an order-of-magnitude estimation, based on the assumption that activation energy is the only barrier to formation of a radiative singlet state.

‖ This assumes a purely diffusion-controlled reaction and would have a rate constant of 10^{10} M^{-1} sec^{-1}.

Interestingly, the emission behaviour was altered by the presence of oxygen only quantitatively, not qualitatively [21].

A description of the electrochemical processes occurring during ECL of these ruthenium chelate complexes, especially the $(Ru(bipy)_3^{2(+)}$ complexes in terms of the corresponding MO diagram has been given [21].

An interesting further development in ECL has been the preparation and investigation of polymer-coated modified electrodes [22]. Special interest has been directed towards the possible catalytic properties of such electrodes. As a model, ECL was investigated with Nafion* – coated electrodes in which a large amount of $Ru(bipy)_3^{2(+)}$ was bound by the usual ionic binding in an ion exchange polymer. Anodic oxidation of the $Ru(bipy)_3^{2(+)}$ to the 3(+)-charged complex in the presence of oxalate ions in aqueous solution yields the intense orange chemiluminescence characteristic of $[Ru(bipy)_3^{2(+)}]^*$. The mechanism for this chemiluminescence has been suggested to be the following [22]:

a) $Ru(bipy)_3^{2(+)} \rightarrow Ru(bipy)_3^{3(+)} + e^{(-)}$
b) $Ru\ (bipy)_3^{3(+)} + C_4O_4^{2(-)} \rightarrow Ru(bipy)\ 2^{(+)} + C_2O_4^{(\pm)}$
c) $C_2O_4^{(\pm)} \rightarrow CO_2 + CO_2^{(\pm)}$
d) $CO_2^{(\pm)} + Ru(bipy)\ 3^{3(+)} \rightarrow CO_2 + [Ru\ (bipy)^{2(+)}]^*$
e) $CO_2^{(\pm)} + Ru\ (bipy)_3^{3(+)} \rightarrow CO_2 + Ru\ (bipy)^{2(+)}$
f) $[Ru\ (bipy)^{2(+)}]^* \rightarrow Ru(bipy)^{2(+)} + h\nu$
g) $[Ru\ (bipy)_3^{2(+)}]^* \rightarrow Ru(bipy)_3^{2(+)}$
h) $CO_2^{(\pm)} + Ru(bipy)_3^{2(+)} \rightarrow CO_2 + Ru(bipy)_3^{(+)}$
i) $Ru\ (bipy)_3^{(+)} + Ru\ (bipy)_3^{3(+)} \rightarrow Ru(biby)_3^{2(+)*} + Ru(bipy)_3^{2(+)}$

The effect of the polymer-coating of the electrode appears to be not only the "immobilisation" of the ruthenium complex but also adsorption of the oxalate ions on the polymer layer. Possible practical applications of such polymer-coated electrodes are in the electrochemical analysis of trace compounds by concentration in the film [24, 25] and electroanalysis of slow reactions, e.g. mediation of enzyme redox processes [22].

In the ECL reaction just described the radical anion acts as a strong reductant. However, in the ECL reaction between peroxydisulfate and $Ru(bipy)_3^{1(+)}$ $Ru(bipy)_3^{1(+)} + SO_4^{(\pm)} \rightarrow [Ru(bipy)_3^{2(+)}]^* + SO_4^{2(-)}$, light is produced by the strong oxidant sulfate radical anion, in acetonitrile/water solutions, according to the reaction sequence:

a) $Ru(bipy)_3^{2(+)} + e^{(-)} \rightarrow Ru(bipy)_3^{1(-)}$
b) $Ru(bipy)_3^{1(-)} + S_2O_8^{2(-)} \rightarrow Ru(bipy)_3^{2(+)} + SO_4^{(\pm)}$
c) $Ru(bipy)_3^{1(+)} + SO_4^{(\pm)} \rightarrow [Ru(bipy)_3^{2(+)}]^* + SO_4^{2(-)}$

By careful choice of the reaction conditions, quenching of $[Ru(bipy)_3^{2((+)}]^*$ by peroxodisulfate dianion $S_2O_8^{2(-)}$ and the problem of $Ru(bipy)_3^{1(-)}$ instability can be minimized, so that a relatively high ECL efficiency (λ_{max} 625 nm) is obtained.

The ECL involving the irreversible oxidation of oxalate is not a regenerative ECL system. It has been shown, however, that it is also possible to produce

---

* Nafion is perfluoro polystyrene sulfonate, a product of E. I. du Pont de Nemours.

polymer-coated electrodes in which ruthenium complex centres are accommodated by appropriate methods of preparation. ECL is produced by pulsing the potential of these coated electrodes at a frequency of 0.5 Hz between $+1,5$ V and $-1,5$ V (vs. SCE). The emission maximum is observed at ca. 610 nm [25].

It appears to be possible to generate ECL also without a supporting electrolyte if thin-layer cells are used. By this technique ECL can be applied for the detection of aromatic hydrocarbons in HPLC [26].

XI.2 Chemiluminescent Reactions of Radical Anions with Various Electron Acceptors

One of the earliest examples of electron transfer leading to chemiluminescence in an organic reaction was reported in 1964 by Chandross and Sonntag [5]. The radical anion of diphenylanthracene (DPA), produced by reaction with potassium metal, was oxidised by 9,10-dichloro-9,10-diphenyl anthracene (DPACl$_2$). Chemiluminescence with a spectrum identical to that of DPA fluorescence was observed.

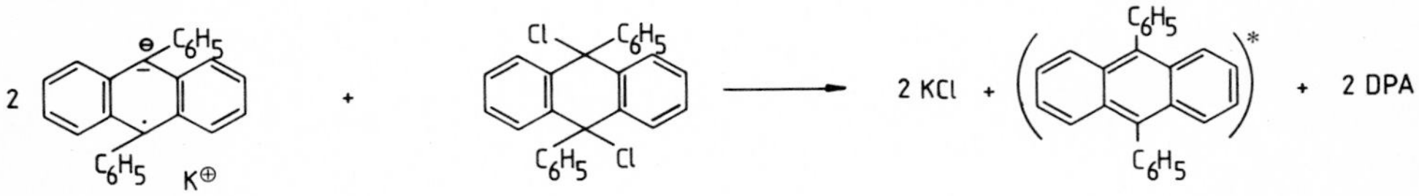

Many other oxidants such as chlorine, HgCl$_2$, oxalyl chloride, p-toluene sulphonyl chloride and benzoyl peroxide gave the same result [5, 27, 28]. The chorine containing compounds may act directly or by formation of DPACl$_2$ to give the reaction already mentioned. Benzoyl peroxide is thought to react with the DPA radical anion forming the benzoyl radical and benzoate anion. The benzoyl radical then abstracts an electron from DPA radical anion to give the first excited singlet state of DPA.

It was pointed out [28] that it was essential to calculate the energy available from the reaction scheme employed, and to show that it was sufficient to populate the excited state observed. In spite of this timely warning of a basic requirement for successful interpretation in the field, papers are still published which do not take account of the energetics of the reaction pathway proposed.

An other example of this very general type is the reaction of naphthalene radical anion with 9-anthranoyl peroxide, the excited state formed being that of anthracene 9-carboxylate.

A very large number of investigations concerning chemiluminescence of chemically produced radical ions has been performed since then.

The radical anions can be prepared by the reaction of a fluorescent aromatic hydrocarbon such as 9,10-dimethylanthracene, with an alkali metal (lithium, sodium or potassium) in an ether solvent (1,2-dimethoxyethane, tetrahydrofuran or 2-methyltetrahydrofuran.

XI.2.1 Wursters's Blue-Type Compounds and Triphenylamines

Stable radical cations are rare but the famous examples known as Wurster's Red and Wurster's Blue have provided a very interesting series of reactions. The first examples of energy deficient chemiluminescence were discovered during this investigation.

They can be obtained by titration of the corresponding p-phenylene diamine with hypobromite solution in the presence of, e.g. sodium perchlorate:

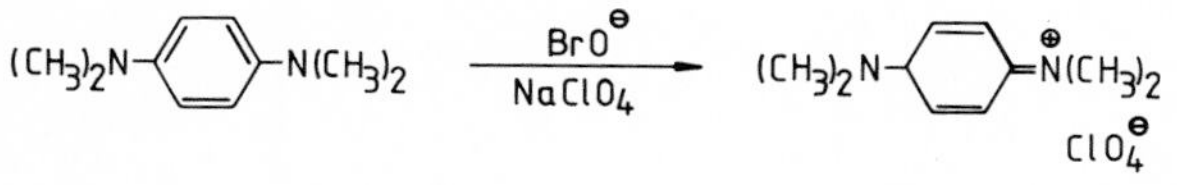

$$\underline{XI}(12)$$

Radical ion recombination was investigated according to the general scheme:

$$2A^{(\pm)} + {}^2D^{(\dot{+})} \rightarrow A + D + h\nu$$

where A is a radical anion (usually an aromatic hydrocarbon derivative) and D the radical cation.

The study of radical ion recombination chemiluminescence of this sort emerged from observations of, so to speak, the reverse reaction. The fluorescence of aromatic hydrocarbons was found to be effectively quenched by certain electron-donor molecules, e.g. N,N-diethylaniline. A new emission appeared in such cases which exhibited a strong red shift, and was ascribed to an excited charge transfer complex ("exciplex", hetero-excimer) (see e.g. [31]), when a nonpolar solvent was used.

$$^1A^* + D \rightarrow \qquad {}^1(A^{(\pm)}D^{(\dot{+})}) \rightarrow A^{(-)} + D^{(+)}$$
Charge-Transfer-Complex

The absorption spectra of such exciplexes could be obtained by laser flash experiments: these spectra resemble those of the respective radical ions [32, 33].

In polar solvents, however, there is no exciplex emission. The "free" radical ions are formed, and they can produce chemiluminescence by annihilation. The recombination reaction between perylene radical anion (14) and tetramethyl-p-phenylene diamine radical cation ("Wurster's Blue") in 1,2-dimethoxyethan (= DME) was the first "bright" chemiluminescence of this type [34].

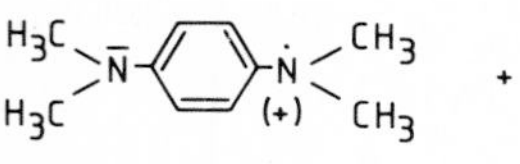

$$\underline{XI}(12)$$
TMPD

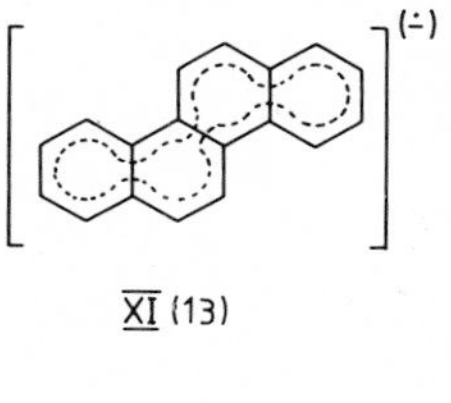

$$\underline{XI}(13)$$

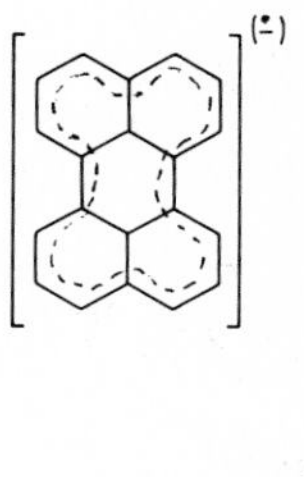

$$\underline{XI}(14)$$

Of course, the same energetic requirements as in the direct production of excited singlet states ("S-route") or the production of the latter via triplet states ("T-route") are valid here as in ECL (see p. 130). As an example, the chemiluminescent reaction of Wurster's Blue radical cation (12) with chrysene radical anion (13) must occur via the "T-route". The emission spectrum matches chrysene fluorescence, but the (12)/(13) redox reaction has an enthalpy corresponding to 2.66 eV only, whereas the energy of the chrysene first excited singlet state is 3.43 eV.

As in energy-deficient systems in ECL, an external magnetic field influences the emission spectrum, as is demonstrated in the Wurster's Blue/4,4'-dimethyl-biphenyl system (Fig. 18).

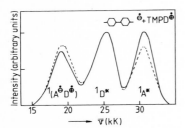

Fig. 18. Chemiluminescence spectra of the 4,4'-dimethylbiphenyl radical anion/tetrame-thyl-p-phenylenediamine radical cation reaction in 1,2-dimethoxyethane: a) without b) with an external magnetic field of ca. 1400 gauss (From Zachariasse [29]).

A large number of aromatic hydrocarbon radical anions has been reacted with Wurster's Blue-type radical cations, e.g. the radical anions of 1-phenylnaph-thalene, 8,8'-dimethyl-naphthalene, 1,1'-binaphthyl, p-terphenyl, chrysene, 1,2-dimethylchrysene [29]. They all represent energy-deficient systems, so that triplet-triplet annihilation had to be regarded as the mechanism for the production of the emitting singlet state. The quantum yields were in the range 10^{-5} to 10^{-4} Einstein/mol [29].

No chemiluminescence at all was observed in the reaction of certain aromatic radical anions with Wurster's Blue, using for example pentacene radical anion, or azulene radical anion. It may be that the redox enthalpy is not sufficient for the production of excited triplet states either of the donor or the acceptor molecule.

Alternatively, the lifetime of a primarily produced triplet state may be too short to achieve T-T-annihilation.

In the case of azulene at least, singlet states are not produced by triplet-triplet annihilation.

Quantum yields about ten times higher were measured when Wurster's Blue-type radical cations were replaced by those of tri-p-substituted triphenylamines. These radical cations are relatively stable [35]. The following triphenylamine derived radical cations were investigated:

R = CH$_3$: tri(p-tolyl)amine (TPTA)
R = N(CH$_3$)$_2$: tri-(p-dimethyl-aminophenyl)amine (TPDA)
R = OCH$_3$: Tri(p-anisyl)amine (TPAA)

Most of these and other chemiluminescent radical ion recombination reactions are energy-deficient. One exception is that of TPTA$^{(+)}$/coronene$^{(\pm)}$ where the redox enthalpy available is sufficient for direct production of the coronene first excited singlet state [29].

Table 11. Radical ion recombination chemiluminescence of aromatic hydrocarbon radical anions with subst. triphenylamin radical cations (after Zachariasse [29])

Oxidant	Reductant (Radical Anion)	Solvent	Chemiluminescence Quantum Yield $\emptyset_{CL}$	ref.
Tri-(p-tolyl)amine-radical cation (= TPTA$^{(+)}$)	9,10-Dimethyl-anthracene	THF	7.5×10^{-2}	[36]
TPTA$^{(+)}$	Anthracene	THF	5.0×10^{-2}	
TPTA$^{(+)}$	Pyrene	THF	3.0×10^{-2}	
TPTA$^{(+)}$	Naphthalene	DME	0.7×10^{-2}	[29]
Tri-(p-dimethyl-aminophenyl)amine radical cation (TPDA$^{(+)}$)	Naphthalene	THF	0.5×10^{-2}	
TPDA$^{(+)}$	9,10-Dimethyl-anthracene	DME	0.4×10^{-3}	
TPTA$^{(+)}$	Fluoranthene	DME	0.2×10^{-2}	

In nearly all radical anion recombination reactions with TPTA$^{(+)}$ exciplex emission is observed – the exciplex being formed directly from the radical ions. Mixed triplet-triplet annihilation appears to be impossible as the triplet lifetime of TPTA is very short [37].

No effect of an external magnetic field was observed in TPTA$^{(+)}$/radical anion reactions.

The quantum yields of radical ion recombinations with p-substituted triphenylamines are distinctly higher than those of systems which produce the excited products via the T-route. It has been proposed that in the former case thermal dissociation of the exciplex produces either $^1A^*$ or $^1D^*$ species [29, 36]. There is a temperature effect on the relation between exciplex and $^1A^*$ emission: lowering the temperature enhances exciplex emission.

Therefore, in addition to the S- and the T-route a third mechanism has been suggested [36], and called the C-route [3].

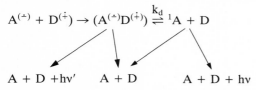

k_d: rate constant of exciplex dissociation (temperature-dependent) For details see [3, 36].

The chemiluminescent reactions between DPA radical anion and chlorine or several inorganic (e.g. HgCl$_2$) and organic (oxalyl chloride, toluenesulfonylchloride) chlorides were suggested as proceeding via the chlorinating action of these compounds, e.g. [39]:

$$DPA^{(-)} + HgCl_2 \longrightarrow DPACl\cdot + \cdot HgCl$$

$$DPA^{(-)} + DPACl\cdot \longrightarrow DPA^* + DPA + Cl^{\ominus}$$

XI.2.2 Electron Transfer Reactions in Metal Complexes

Although the study of ruthenium ECL (Sect. X.1) is convenient, the first experiments were performed not electrochemically, but with hydrazine or hydroxyl ion as electron-transferring agents. The general mechanism [40, 41] is:

$$\text{Metal (ligand)}_x^{n+1(+)} + e^{(-)} \to \text{metal (ligand)}_x^{n(+)} + h\nu$$

Ruthenium chelates of 2,2'-bipyridyl (15), 5-methyl-o-phenanthroline (16) and the similar compounds (17) and (18) were investigated.

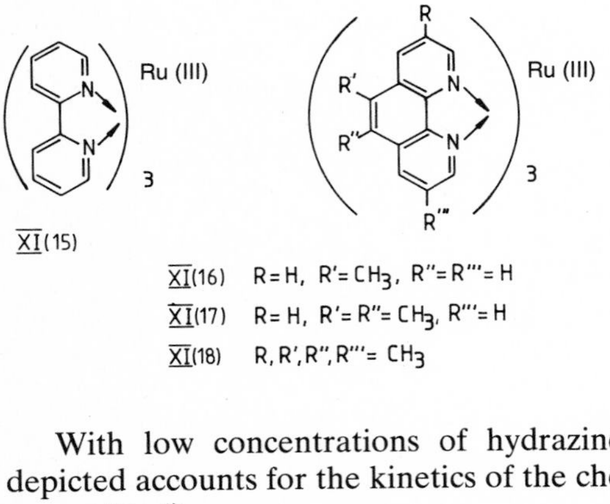

$\overline{XI}(15)$

$\overline{XI}(16)$ R = H, R' = CH$_3$, R'' = R''' = H
$\overline{XI}(17)$ R = H, R' = R'' = CH$_3$, R''' = H
$\overline{XI}(18)$ R, R', R'', R''' = CH$_3$

With low concentrations of hydrazine as reductant, the mechanism as depicted accounts for the kinetics of the chemiluminescent reaction (the ligands are omitted):

Ru(III) + N$_2$H$_4$ $\quad\quad$ → Ru(II)* + N$_2$H$_3$· (One-electron-transfer from hydrazine to Ru(III))

2N$_2$H$_3$· $\quad\quad\quad\quad$ → N$_2$ + 2 NH$_3$
Ru(III) + N$_2$H$_3$· $\quad$ → HN=NH + Ru(II) + H$^{(+)}$
Ru(III) + HN=NH $\quad$ → Ru(II)*
Ru(II)* $\quad\quad\quad\quad$ → Ru(II) + hν

This reaction as written is energy-deficient, i.e. the first excited singlet state of Ru(II) is inaccessible directly. The emission of the Ru(II) species, produced in this electron-transfer reaction, was therefore ascribed to a charge transfer, spin-forbidden luminescence [41].

The simplest means of reduction of ruthenium which yields chemiluminescence is by the use of solvated electrons [30]. $Ru(bipy)_3Cl_2$ solutions were oxidized by PbO_2 to the corresponding Ru(III) compound which was treated in a solution of water, t-butyl alcohol and acetate buffer by 4–40 nsec single pulses of 15 MeV electrons. The chemiluminescence obtained by this experiment is seen in Fig. 19 [30]:

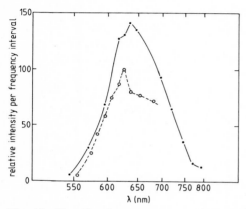

Fig. 19. Chemiluminescence spectrum of $Ru(bipy)_3^{3(+)}$ reacting with $e_{aq}^{(-)}$ [30]

The intensity of the $Ru(bipy)_3^{3(+)}$ reaction with hydrazine as a function of time could be simulated by a computer programme based on the mechanism formulated on p. 144.

XI.3 Distinctions between Radical Mechanisms for the Generation of Excited States

Since the minimum description of an excited state involves the separation of electrons into bonding and anti-bonding orbitals, radical processes have often featured in the explanations. However it is important to distinguish the various mechanisms since some are very much better understood than others. The most satisfactory mechanism to date is that named CIEEL by Schuster and anticipated by McCapra in his explanation for the chemiluminescence of the active oxalates. The essence of the chemically initiated electron exchange luminescence process is that there are two electron transfer steps. These steps take place within the first encounter complex formed from the two reacting molecules. The first transfer is strongly dependent on the difference in redox potential between the oxidant and reductant. This dependency is seen in the need for fluorescent molecules of low ionisation potential. The oxidant, usually a peroxide, similarly often has a particular structural requirement. Thus simple alkyl peroxides rarely if ever act as suitable oxidants in this mechanism. The transformation of the intermediate alkyl peroxide in the reaction path of the luciferins (p. 152, 154) to a strained perester – a considerably better oxidant – is a good example.

Once electron transfer has occurred, coulombic attractions keep the oppositely charged ions in the solvent cage. If there is an accessible excited state of one of the partners, then the back transfer of the electron will occur at an immeasurably fast rate. This makes direct examination of this crucial part of the mechanism extremely difficult, if not impossible. Nevertheless, there is good indirect evidence for this sequence of events. The remaining essential feature of the mechanism is that in order to allow the formation of an energetic excited state within the second law of thermodynamics, the second electron transfer must occur from the now reduced oxidant so as to leave a stable molecule. Certain structural types are thus required. Examples of these can be seen in the reactions depicted on pages 47 and 35.

In the case of pre-formed radical anions, whether produced chemically or electrochemically, all that is necessary for excited state production is that the oxidant should be capable of abstracting an electron with release of sufficient energy to satisfy the thermodynamic requirements.

The other radical mechanisms are much less well defined. The problem they present is typified by the common observation that almost any mixture of fluorescent or potentially fluorescent compounds and powerful oxidants such as peroxides or ozone will produce measurable and indeed sometimes easily visible light emission. Some of these mechanisms will, by presently obscure and circuitous routes, produce intermediates which can react by the CIEEL path. This may be the case for the Grignard reaction (p. 28). However there are cases such as the reaction between per-acids and fluorescers (p. 42) for which no mechanism can be envisaged as yet.

Another class of radical is considered at some length elsewhere in this book. The chemiluminescent decomposition of alkyl peroxides occurs by a variety of routes, but the invariable formation of a carbonyl group in an excited state removes much of the difficulty attending the examples discussed immediately above. There are some well understood cases such as the dioxetans and a range of more obscure ones such as those involved in luminescence attending phagocytosis and polymer degradation.

Lastly, the simplest radical-based luminescence of all has been investigated [42]. The energy available from the bond formed by the recombination of two carbon based radicals releases sufficient energy to populate a suitable excited state. This is far from being a general reaction, but further investigation may prove fruitful.

XI.4 References

1. Hercules, D. M., in Physical Methods of Organic Chemistry, Part II, p. 257, Weissberger, A. and Rossiter, B. (Eds.), Academic Press, New York 1971

2. Gundermann, K.-D., Chemilumineszenz organischer Verbindungen, p. 120, Springer-Verlag, Berlin 1968
3. Faulkner, L. R., Int. Rev., Sci.: Phys. Chem. Ser. *2*, 213 (1976)
4. Faulkner, L. R., in Methods in Enzymology, 2 VII, DeLuca, M. (ed.), 404
5. Chandross, E. A. and Sonntag, F. I., J. Amer. Chem. Soc. *86*, 5350 (1964)
6. Hoytink, G. J., Discuss. Farad. Soc. *45*, 14 (1968)
7. Maricle, D. L. and Maurer, A., J. Amer. Chem. Soc. *89*, 189 (1967)
8. Maloy, J. T., Prater, K. B. and Bard, A. J., ibid. *93*, 5959 (1971)
9. Frumkin, A. N., Nekrasov, L., Levich, V. G. and Ivanov, Y. B., J. Electroanal. Chem. *1*, 84 (1959)
10. Maloy, J. T. and Bard, A. J., J. Amer. Chem. Soc. *93*, 5968 (1971)
11. Zweig, A., Maurer, A. H. and Roberts, B. G., J. Org. Chem. *32*, 1322 (1967)
12. Weller, A. and Zachariasse, K., Chem. Phys. Lett. *10*, 424 (1971)
13. Bard, A. J., Keszthelyi, C. P., Tachikawa, H. and Tokel, N. E., in: Cormier, M. J., Hercules, D. M. and Lee, J. (Eds.), Chemiluminescence and Bioluminescence, p. 193, Plenum Press, New York 1973
14. Bezman, R. and Faulkner, L. R., J. Amer. Chem. Soc. *94*, 3699 (1972)
15. Porter, G. and Wilkinson, F., Proc. Roy. Soc. (London), Ser. A., *264*, 1 (1961)
16. Freed, D. J. and Faulkner, L. R., J. Amer. Chem. Soc. *93*, 2097 (1971)
17. Haas, J. W. and Baird, J. E., Nature (London) *214*, 1006 (1966)
18. Bezman, R. and Faulkner, L. R., J. Amer. Chem. Soc. *95*, 3083 (1973)
19. Lytle, F. E. and Hercules, D. M., Photochem. Photobiol. *13*, 123 (1971)
20. Tokel, N. F. and Bard, A. J., J. Amer. Chem. Soc. *94*, 2862 (1972)
21. Tokel-Takvoryan, N. E., Hemingway, R. E. and Bard, A. J., ibid. *95*, 6852 (1973)
22. Rubinstein, I. and Bard, A. C., ibid. *103*, 5007 (1981) and references cited
23. Oyama, N. and Anson, F. C., J. Electrochem. Soc. *127*, 247 (1980)
24. Oyama, N., Shimomura, T., Shigehara, K. and Anson, F. C., J. Electroanal. Chem. *112*, 271 (1980)
25. Abruna, H. D. and Bard, A. J., J. Amer. Chem. Soc. *104*, 2641 (1982)
26. Schaper, H., Köstlin, H. and Schnedler, F., GIT Fachz. f. d. Lab. *26*, 745 (1982)
27. Visco, R. E. and Chandross, E. A., J. Amer. Chem. Soc. *86*, 5350 (1964)
28. Hercules, D. M., Science *145*, 808 (1964)
29. Zachariasse, K. A., Thesis, Vrije Universiteit Amsterdam 1972
30. Martin, J. F., Hart, E. J., Adamson, A. W., Gafney, H. and Halpern, H., J. Amer. Chem. Soc. *94*, 9238 (1972)
30 a. Siegel, T. M. and Mark jr., H. B., ibid: *94*, 9020 (1972)
31. Leonhardt, H. and Weller, A., Ber. Bunsenges. Physik. Chem. *67*, 791 (1963)
32. Potashnik, R., Goldschmidt, C. R., Ottolenghi, M. and Weller, A., J. Phys. Chem. *75*, 1025 (1971)
33. Schomberg, H., Staerk, H. and Weller, A., Chem. Phys. Lett. *21*, 433 (1973)
34. Weller, A. and Zachariasse, K., J. Chem. Phys. *46*, 4984 (1967)
35. Walter, R. I., J. Amer. Chem. Soc. *77*, 5999 (1955)
36. Weller, A. and Zachariasse, K. A., Chem. Phys. Lett. *10*, 590 (1971)
37. Förster, F., Dissertation, Univ. Stuttgart 1971
38. Rapaport, F., Cass, M. W. and White, E. H., J. Amer. Chem. Soc. *94*, 3168 (1972)
39. Chandross, F. A. and Sonntag, F. I., ibid. *88*, 1089 (1966)
40. Hercules, D. M. and Lytle, E. F., ibid. *88*, 4745 (1966)
41. Lytle, F. E. and Hercules, D. M., ibid. *91*, 253 (1969)
42. Abuin, E. and Lissi, E., Photochem. Photobiol. *30*, 59 (1979)

C. Bioluminescence

XII. Bioluminescence as a Special Case of Chemiluminescence

Interest in living examples of luminescence is of course older than the study of chemiluminescence. The organic chemist became interested in both in a systematic way about twenty five years ago. It could be argued that investigation of the more complex phenomenon of bioluminescence has in fact been of greater value in understanding the chemiluminescence of synthetic compounds than the reverse. There are several reviews of the biological [1, 2], biochemical [3, 4] and chemical aspects [5, 6, 7] of bioluminescence and only the briefest outline of this delightful natural manifestation will be made. However, the chemiluminescence of model compounds has played a large part in explaining the details of the all – important excitation step, and this aspect is emphasised here.

XII.1 Bioluminescent Organisms

There are many thousands of different species of luminescent organisms. Some are familiar, such as the firefly (and the larval form, the glowworm), lantern fish and the microscopic plankton responsible for the so-called "phosphorescence" of the sea. The great majority of organisms however inhabit the seas, many of them in the deepest parts. So widespread is the occurrence that it has been estimated that greater than 70% of all organisms below 400 m are luminescent [8].

The level of understanding of the chemistry behind the emission of light varies enormously for each organism. In many cases extensive investigation has failed to identify any molecule on which to base a mechanism. Accordingly only the light systems of those organisms for which chemical structures have been assigned are mentioned, the reader being referred elsewhere [1, 2] for the large number of intriguing unsolved problems presented by some remarkable animals. The names of the components were coined in the late nineteenth century by Dubois [9] during his investigation of the boring clam Pholas dactylus (in which no chemical structure has been identified) and the click beetle Pyrophorus, which has a chemical system closely related to that of the well understood firefly.

Dubois extracted the luminous parts of these organisms with both hot and cold water. The hot water extract denatured the enzyme, which he called a luciferase but contained the substrate, called a luciferin. The cold water extract, by permitting enzymic reaction exhausted the luciferin, but contained active enzyme. Mixing of hot and cold extracts re-created the light reaction. Not all luminous organisms lend themselves to this crude assay procedure, since they often require co-factors, or indeed intact cells for the complete light reaction. Nevertheless it presents the simplest biochemical description of the light emitting event.

$$\text{Luciferin} + \text{Luciferase} \xrightarrow{O_2} \text{Oxyluciferin*}$$

$$\text{Oxyluciferin*} \rightarrow \text{Oxyluciferin} + \text{light}$$

The modern term oxyluciferin has been added for clarity, together with recognition of the fact that oxygen is always required in some form. The excited state of oxyluciferin is formed in good yield in most cases, and the light emitted is that of its fluorescence. It should be noted at this stage that this fluorescence is sometimes not that of the free molecule in solution, but is modified by being bound to the enzyme. An outline of the accepted, known mechanisms will be given, followed by a discussion of model reactions which have assisted in arriving at the various explanations. These model reactions often preceded detailed investigation of the enzyme catalysed reactions.

XII.1.1 The Firefly

The first luciferin for which a structure was obtained, and synthesis achieved, was that of the North American firefly, Photinus pyralis [10].

As far as is known, all other fireflies use the same compound in essentially the same way. A notable feature is the requirement for ATP, the result being the formation of luciferyl adenylate (1). This reaction predisposes the luciferin to attack by base and oxygen, and provides the leaving group which results in the formation of the dioxetanone (2).

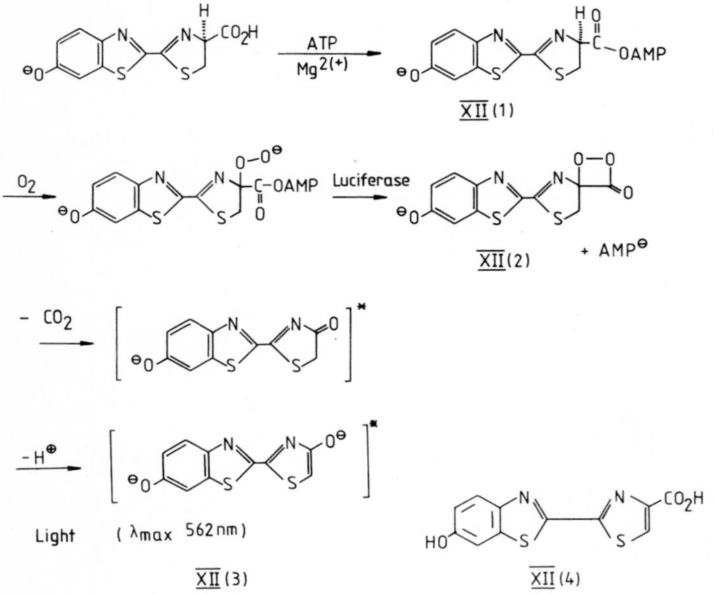

$$\underline{XII}(1)$$

$$\underline{XII}(2) \qquad + \text{AMP}^{\ominus}$$

Light (λ_{max} 562 nm)

$$\underline{XII}(3) \qquad\qquad \underline{XII}(4)$$

Other interesting points were first indicated by model reactions, then later substantiated by direct examination of the reaction on the enzyme. These include the fact that the phenoxide ion is essential for high light yields (suggesting electron transfer – see later). The second ionisation to produce the species (3) which radiates in vivo was also implicated by model studies [11].

XII.1.2 Models for Firefly Luciferin

Although it had been known that the difficult to prepare luciferyl adenylate was chemiluminescent in DMSO and strong base, a problem is caused by the competing dehydrogenation to dehydroluciferin (4). Indeed dehydroluciferin was once considered to be the excited product [12]. The synthesis of a molecule with blocking gem-dimethyl groups (5,R = AMP-adenosyl monophosphate – and R = Ph) showed unequivocally [13, 14] that the excited product was the carbonyl compound (6). However the light emitted was a brilliant deep red (λ_{max} about 600 nm).

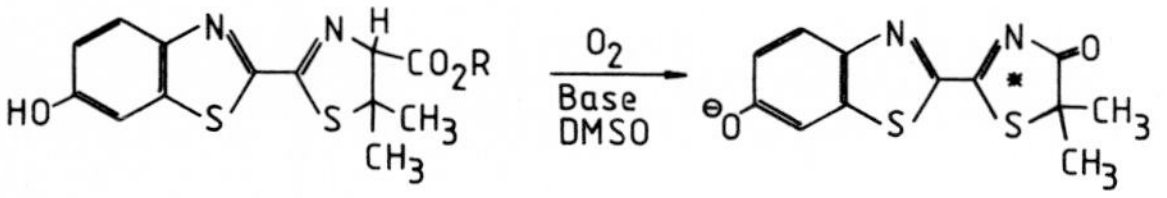

XII (5)   XII (6)

White [11] showed that only if ionisation (enolisation) in the thiazoline portion of the luciferin and its analogues were possible did the more typical yellow light result. If the active esters of (7) and (8) are oxidised in basic DMSO, yellow light predominates at higher base concentrations confirming that deprotonation in the newly formed excited state results in an enolate fluorescing around the typical 562 nm region.

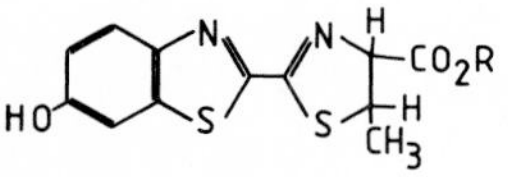

XII (7) 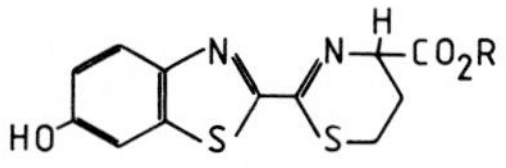   XII (8)

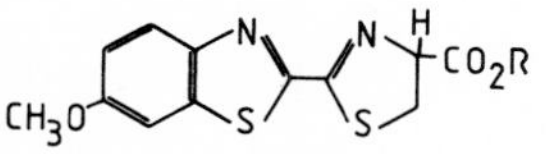

XII (9)

The methyl ether (9) as an active ester gives virtually no light under the same conditions, indicating the need for the strongly electron donating phenoxide ion. Active esters for these chemiluminescence experiments are essential, but rather difficult to synthesise. The only reported examples are

R = AMP, R = Ph [14] and R = C(OEt)=CH$_2$ [15]

the latter being the most convenient, requiring only mixing with ethoxyacetylene.

These models are all particularly accurate indicators of the luciferase-luciferin-O_2 reaction since spontaneous oxidation of the carbanion formed by base seems to be occurring in both the chemical and biochemical systems. In addition the quantum yield of excitation using the ethoxyvinyl ester is about 30%, indicating the power and significance of model reactions.

XII.1.3 Cypridina [16] and Coelenterate [5, 17] Bioluminescence

Although these organisms are not related (the first is an ostracod crustacean and the second is a group of organisms of which the jellyfish is a member), they share the same chemiluminescence mechanism. The luciferins are not identical but the central imidazopyrazine nucleus is common to both. Indeed the most likely biosynthesis for each is the cyclisation of a tripeptide, so that only the amino acid side chains are different. Coelenterate luciferin is also known as coelenterazine and is

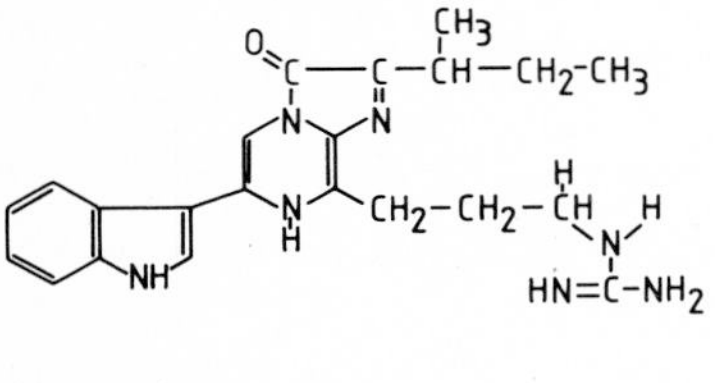

Cypridina Luciferin

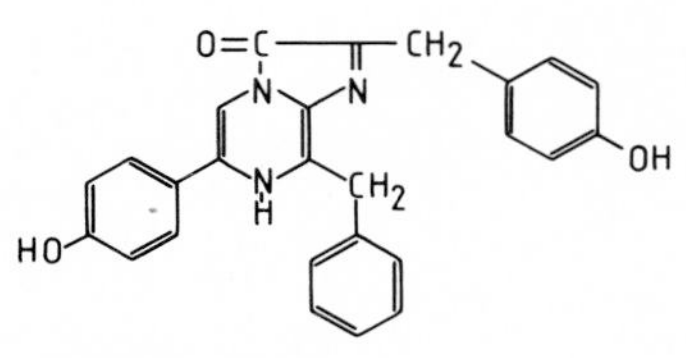

Coelenterate Luciferin

the most ubiquitous of all luciferins, being found in organisms as different as hydroids, fish and squid [18].

Although chemiluminescent experiments can be carried out on the luciferins themselves, synthesis is not easy, so that models in which the central nucleus is more simply substituted make very effective chemiluminescent compounds.

The first model compound [19] to be synthesised (see scheme) shared all the properties of the luciferins – most noticeably its acidity (pK$_a$ 8.3), forming a yellow anion in DMSO/base. This anion oxidised spontaneously and the mechanism shown predicted in all important respects that now accepted for the natural system. The blue emission spectrum was

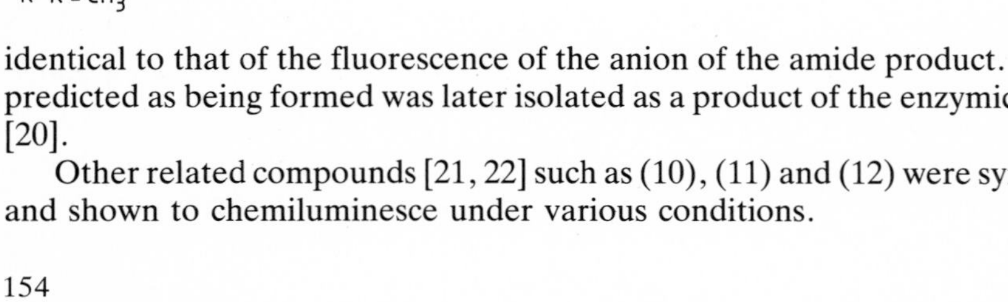

$R^1 = C_6H_5$
$R^2 R^3 = CH_3$

identical to that of the fluorescence of the anion of the amide product. The CO_2 predicted as being formed was later isolated as a product of the enzymic reaction [20].

Other related compounds [21, 22] such as (10), (11) and (12) were synthesised and shown to chemiluminesce under various conditions.

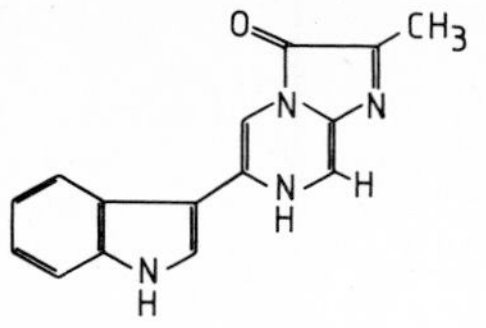

XII (10) 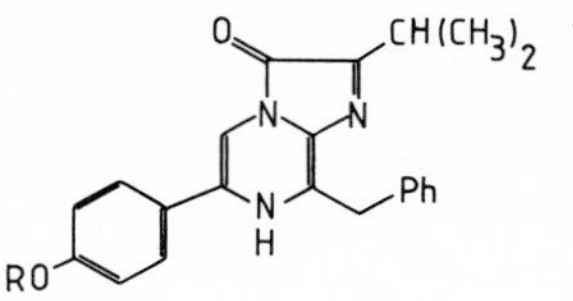 XII (11) 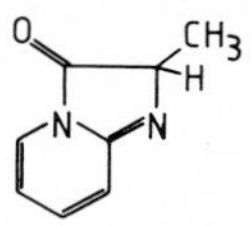 XII (12)

Note that a pyridine ring may be substituted for the pyrazine, with little effect on properties. During these studies some insight was obtained into the lifetimes of the various species formed in excited states. For example, using sodium acetate as base in diglyme [21], it could be shown that the excited state first formed in the oxidation of (10) was that of the anion, and that protonation occurred within the lifetime of this excited state. Two groups of workers also showed [23, 24] that where the group R in (11) was alkyl, similar behaviour was obtained. With R = H we have an excellent model for coelenterate luciferin. Organisms which possess this luciferin show a wide range of emission maxima, and the model duplicates some of these. (Others are the result of energy transfer and are not discussed here.) When (11, R = H) is oxidised in media of various basicities, it can be shown that the amide *anion* is formed in the excited state even when the medium is insufficiently basic to deprotonate the very much more acidic phenol. This again shows that the excited anion is formed *not as a result of proton removal* but *directly* in the excited state. Thus an extra piece of information is available which requires that an excited anion is formed under all conditions. This result is best explained by the last, excitation step being an electron transfer as shown below.

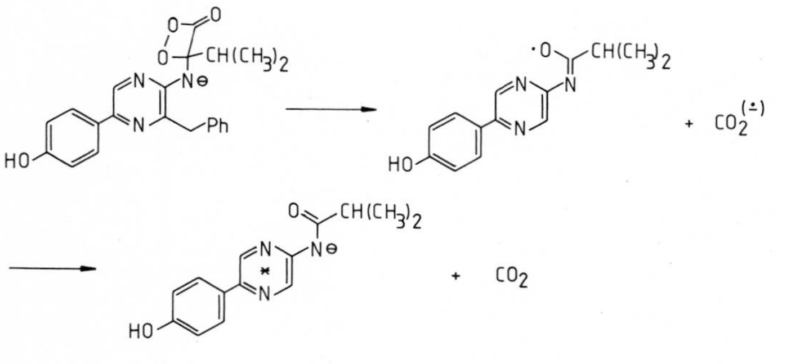

Oxidation of the precursor anion in these models can be catalysed by cobalt (and other transition metals) [25]. The enzyme can also be modelled to some extent by surfactant solutions [26]. Quaternary ammonium salts (e. g. cetyl trimethylammonium bromide) are better than neutral detergents. Anionic surfactants are ineffective. The reason for the activity is two-fold. The rate of oxidation is increased, and the fluorescence of the product, which is very weak in pure water, is considerably enhanced in the hydrophobic micelles.

XII.1.4 Bacterial Bioluminescence

Luminous bacteria [3] are found in almost all the oceans of the world, both free living and as symbionts. They are saprophytes and may be cultured from most scraps of dead marine animals washed ashore. In symbiosis with certain fish they provide continuous light from special organs provided by the host. These organs are often under the control of the host, which can extinguish the light by closing an opaque membrane over the organ, as in the fish Photoblepharon.

The biochemical picture which has emerged largely as a result of work by Hastings [3, 27] and his co-workers is shown below

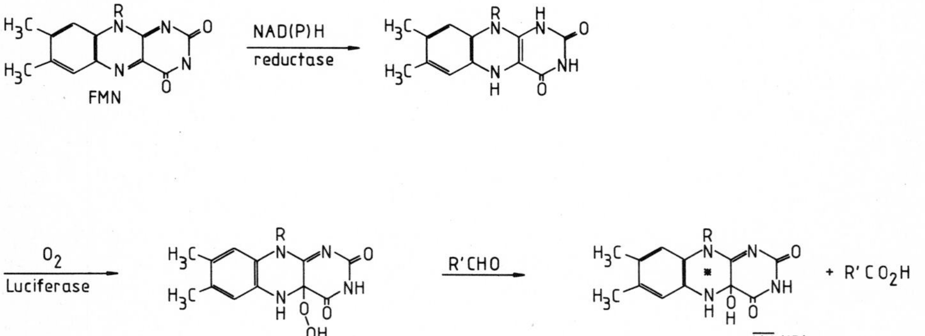

R= ribityl   R′= C_nH_{2n+1}

Although there is some difficulty in achieving complete acceptance of the pseudo-base (13) as the excited product, it must be the most likely candidate at present. It will subsequently dehydrate to re-form FMN as required by the experimental evidence.

The most reasonable intermediate formed from the long chain aldehyde (R′ is most often $CH_3(CH_2)_{10}$) is (14)

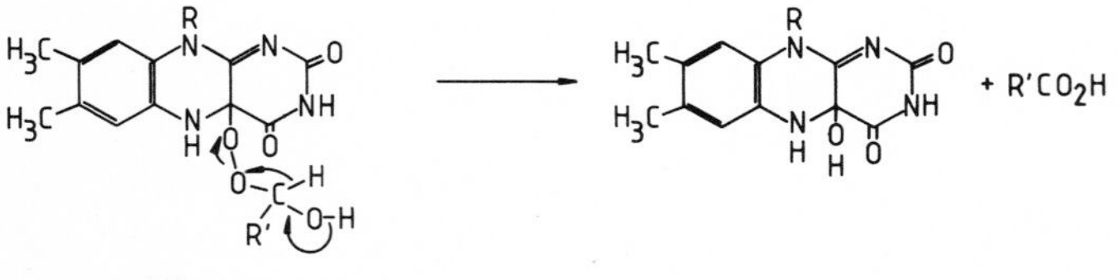

XII (14)

A mechanism related to that of the Baeyer-Villiger reaction (as shown) is highly likely, and is confirmed by a substantial isotope effect for the removal of the aldehydic H-atom [29].

Very severe problems are encountered in providing a set of model reactions which might explain why the Baeyer-Villiger-type decomposition is luminescent.

156

As far as is known, no examples of luminescent reactions of this sort have been discovered.

Addition of various peroxides in buffer to flavinium salts such as (15) are presumed [30] to have given an intermediate related to the key compound in the bacterial reaction.

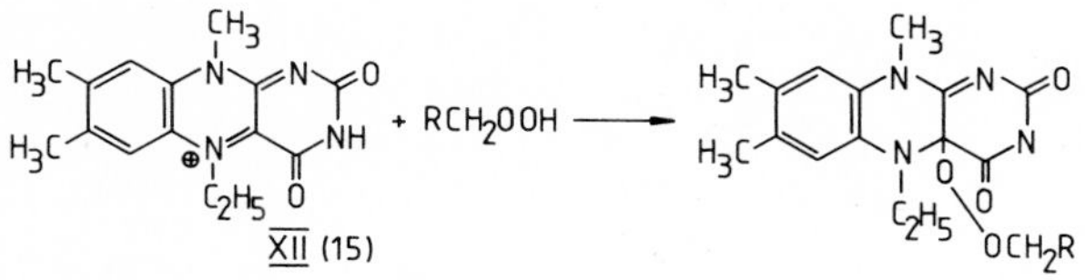

The reaction was accompanied by light emission, but subsequent work [31] showed that this low level light bore no relationship to any reasonable reaction scheme. It was most likely to have arisen from generalised peroxide decomposition, taking place in the presence of a variety of fluorescent by-products. The pseudo-base previously mentioned is only fluorescent when enzyme bound. Thus the model just discussed was unlikely to realise any identifiable excitation pathway, since the end product would be non-fluorescent and therefore incapable of proving the point.

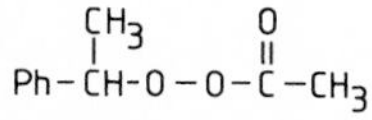

Secondary peroxyesters [32] such as (16) are the closest analogues of the presumed intermediate, but are still inadequate models since the flavin peroxy-adduct is not an ester. Such esters have to be heated to achieve reaction, and a good model should show similar characteristics and not require forcing conditions.

An example of a much closer structural analogy was devised by Richardson [33]. The scheme below shows the reaction sequence, but even with energy acceptors such as DBA and rubrene the excitation yield was quite insignificant with $\emptyset_{CL} = 10^{-10}$.

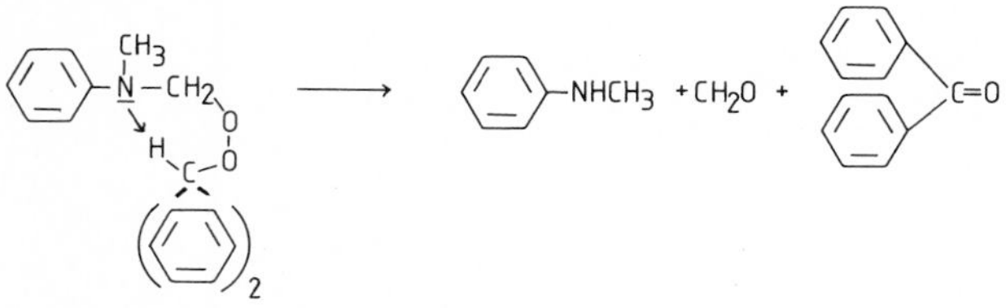

Thus in every case so far model systems have not duplicated the bacterial system. Insofar as they are valid in indicating reaction pathways (given the enormous success in the two cases previously discussed), it may be necessary to re-interpret the biochemical findings in the light of the negative evidence.

XII.2 General Models for Bioluminescent Reactions

The reactions just discussed use compounds more or less closely related to those involved *in vivo*. In order to identify the general reaction type and structural features operating, it is helpful to consider less closely related structures. For example, the influence of the ionisation of the phenolic hydroxy-group (p. 153) on the efficiency of firefly luminescence is highlighted by the work of Schaap and Gagnon [34] on the dioxetan (17).

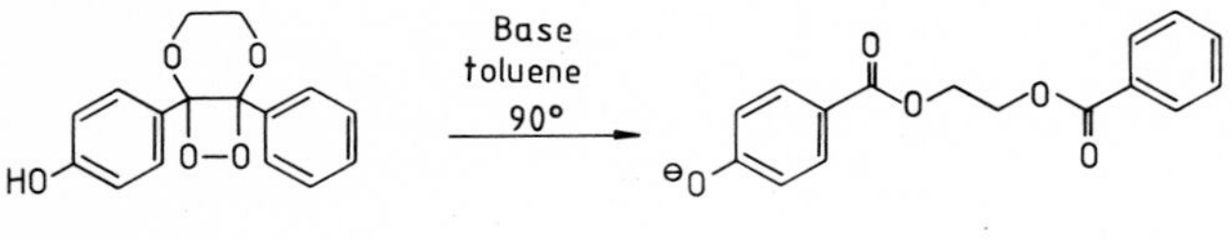

XII (17)

Before ionisation the dioxetan behaves as a simple alkyl substituted member of the series, giving very low singlet excitation yields and being moderately stable. Addition of base lowers the stability dramatically – an increase in rate of 4×10^6 times is observed. The yield of singlet excited states increases by over 100-fold, and its lifetime of 46 ms is very similar to the duration of the firefly flash. Thus the increased availability of electrons predisposes even a dioxetan (rather than the more electronegative dioxetanone of the luciferins) to decomposition with formation of singlet excited states.

XII.2.1 Electron-rich Dioxetans

A related phenomenon is seen in the easy decomposition and extraordinarily high quantum yield ($\varnothing_{CL} = 0.13$) of the electron rich dioxetan (18); [35] see also p. 63).

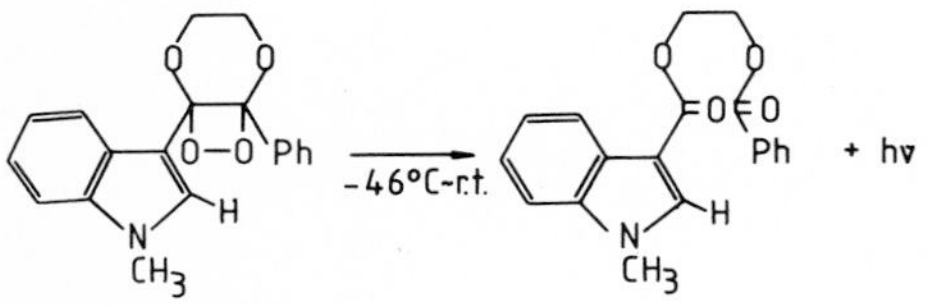

XII (18)

Here solvent effects confirm the electron transfer or charge transfer character of the decomposition. Other electron rich dioxetans are amino-substituted [36], such as (19) and (20)

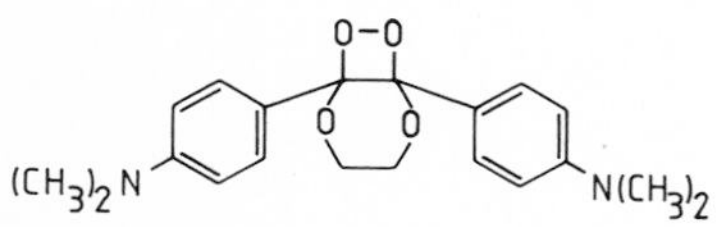

XII (19) 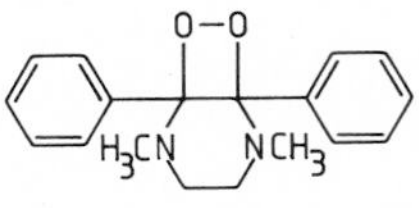   XII (20)

Such dioxetans are often extremely unstable, (20) for example only being stable below $-60°$. They do not always produce high yields of singlet excited states, probably since accessible states are not always available. The reasons for their reactivity are difficult to investigate because of this instability.

Thus the electron donating properties of the essential phenoxide ion in firefly luciferin, and the adjacent N-atoms of coelenterate and Cypridina luciferins, are seen to be important in explaining both the reaction rate and the excitation mechanism.

XII.2.2 Acridans as Model Compounds

Chemiluminescent compounds can often be classified under several of the headings used in this book. Acridan carboxylate esters exemplify some of the properties discussed elsewhere, not least those related to the chemistry of certain luciferins. It is possible to connect autoxidation, peroxide, dioxetan, dioxetanone and electron transfer chemiluminescence in a single series of compounds. A particular advantage is that important intermediates can be characterised and substitution changes readily made in compounds with very high light yields [6, 7]. The reaction proceeds best in dipolar aprotic solvents, and autoxidation of the carbanion with the emission of light occurs on the addition of strong base.

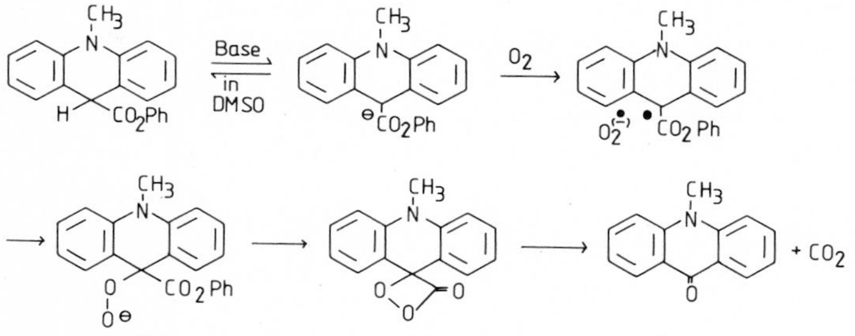

The parallels between the reactions of firefly luciferin, as presently understood, and those of the acridan are striking, as the comparison below shows. Note that the oxidation to form the peroxide does not take place in one step. Investigation [37] has shown that electron transfer and recombination within the solvent cage, very similar to other carbanion oxidations, gives the appearance of a one step reaction.

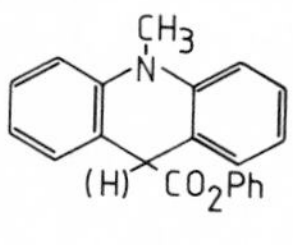

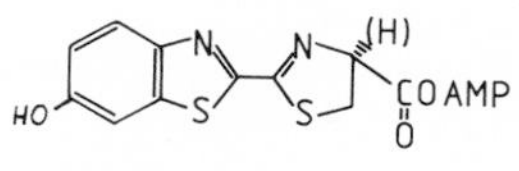

	Acridan ester	Firefly luciferin
pK_a (H)	$20 \pm 0,2$	~ 20
pK_a(leaving group)	$8-10$	~ 6
K_H/K_D	$3,4$	$3,0$
Solvent	DMSO	Hydrophobic site

159

Formation of the carbanion is partially rate determining as is reaction with O_2. The effect of the leaving group and the formation of the dioxetanone can be studied in detail since the peroxide (21) is isolable. Addition of H_2O_2 in base to the acridinium salts can be used to make this all-important intermediate.

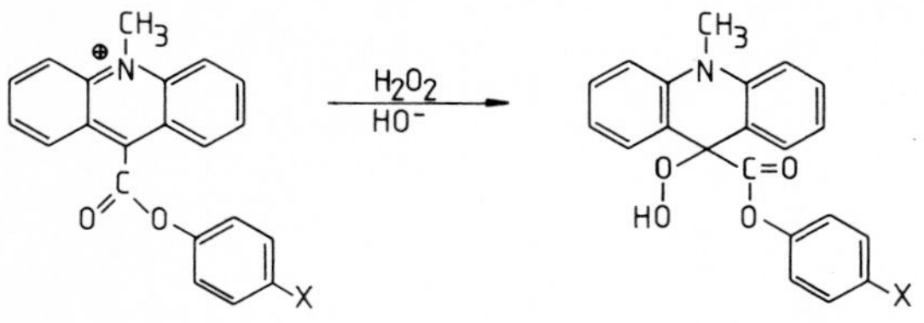

$$\underline{XII} \ (21)$$

XII.2.3 Models for the Excitation Step

The structure giving rise directly to the excited state is now clearly seen as a dioxetanone such as (2) or (22). Cyclisation of the alkyl hydroperoxide forms a strained per-ester – a very much stronger oxidant. Since the electron rich donor is part of the same molecule it is not surprising that reaction is thought to be instantaneous, precluding isolation.

Reaction between a separate dioxetanone and donor molecule (p. 39) shows [38] the strongly catalytic effect derived from the low ionisation potential of the donor.

Dioxetanones such as (23) and (24) and donors such as rubrene

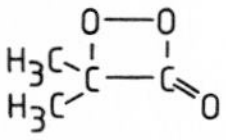

$$\underline{XII} \ (23) \qquad \underline{XII} \ (24)$$

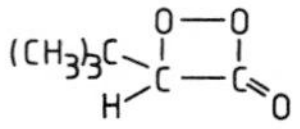

and dihydrophenazines can be considered models for the all important excitation step. The current view is that this proceeds by the CIEEL (electron transfer) mechanism. However although discrete electron transfer may be the mechanism for the intermolecular reaction, more subtle possibilities exist for the intramolecular case. Study of dioxetans and dioxetanones relevant to the luciferins is clearly an important activity.

Dioxetanones are too unstable ever to be isolable when directly attached to the fluorescent donor, but the related dioxetans can usually be characterised. Some have already been discussed (p. 63) but direct relation to the acridan esters would provide the best connection with the luciferins. Even dioxetans present stability problems. For example, treatment of (25) with singlet oxygen gave a dioxetan [39] with a half life of 8 min. at 25 °C, decomposing with a quantum yield of 0.25. This decomposition could be followed by N.M.R.

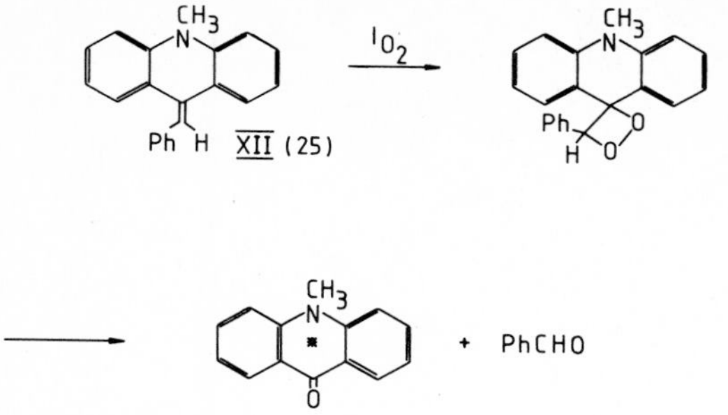

Dioxetans of this type can be made still more stable by the introduction of bulky substituents. The adamantyl derivative (26) can be easily crystallised [40], since it has a distinctly higher activation energy for decomposition (110.5 kJ mol^{-1}, as against 82.7 kJ mol^{-1}). The yield of excited singlet N-methyl acridone is 12%.

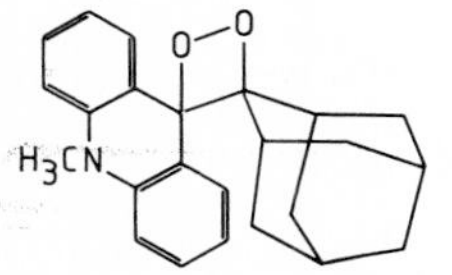

XII (26)

It is possible to write an electron transfer mechanism from the acridan portion to the unoccupied orbital of the O–O bond. However examination of models shows that the orbital overlap is poor, and it seems equally acceptable to see the decomposition as occurring through a mixing of charge transfer states. Such states would have exactly the same requirements as discrete electron transfer, but would better explain the high singlet yields (the statistical probability in totally free electron exchange for singlet formation is only 1 in 4). Thus the final details of the firefly luciferin excitation mechanism would have taken one of two pathways, with preference for the second.

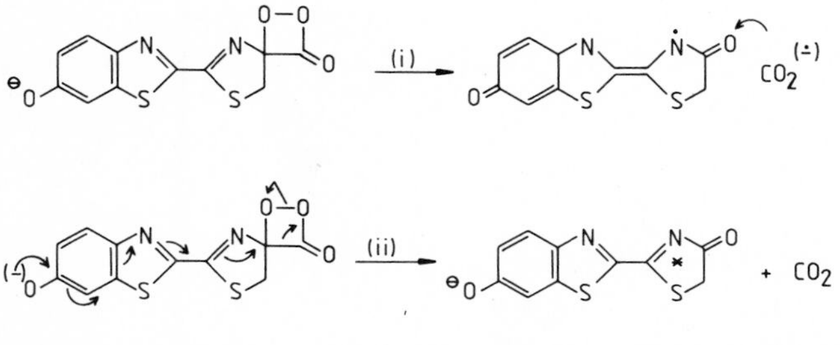

Very similar ideas can be applied to Cypridina and coelenterate luciferin, but excitation in the bacterial system seems at present less understandable.

161

XII.3 Other Bioluminescent Organisms

It is a happy coincidence that the first luciferins whose structures were elucidated have a chemistry which leads to a satisfactory explanation of mechanism. However there are many other organisms with complex biochemical systems, and even where the structures of some of the luciferins are known, no chemiluminescence mechanism has yet been derived. Examples of some of the structures which await further investigation are shown.

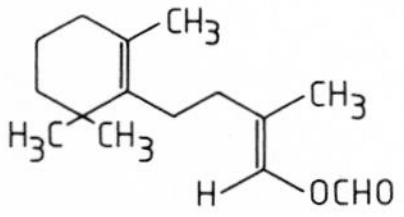

Latia Luciferin [41]

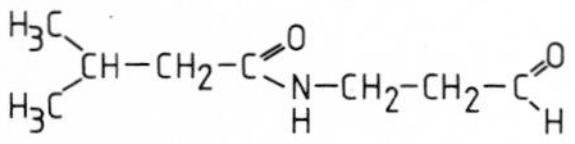

Diplocardia Luciferin [41]

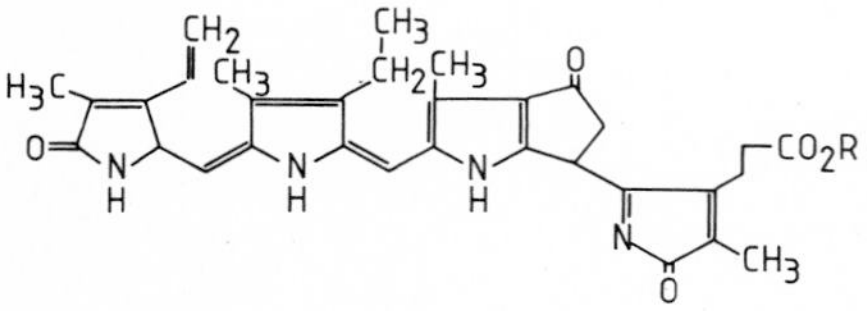

Dinoflagellate Luciferin[43]

XII.4 References

1. Harvey, E. N., Bioluminescence. Academic Press, New York 182
2. Herring, P. J. (ed.), Bioluminescence in Action. Academic Press, London 1978
3. Hastings, J. W. and Nealson, K. H., A. Rev. Microbiol. *31*, 549 (1977)
4. DeLuca, M. A. (ed.), Bioluminescence and Chemiluminescence, Methods in Enzymology *57*, (1978)
5. Cormier, M. J., Wamsler, J. E. and Hari, K., Progress in the Chemistry of Organic Natural Products (Herz, W., Griseback, H. and Kirby, G. W., eds.) *30*, 1, (1973)
6. McCapra, F., Acc. Chem. Res. *9*, 201 (1976)
7. McCapra, F., Proc. R. Roc. Lond. B *215*, 247 (1982)
8. Badcock, J. R. and Merrett, N. R., Prog. Oceanogr. *7*, 3 (1976)
9. Dubois, R., C. r. Séanc. Soc. Biol. Paris (ser. 8) *2*, 559 (1885). Dubois, R., C. r. Séanc. Soc. Biol. Fr. *39*, 564 (1887)
10. White, E. H., McCapra, F. and Field, G. F., J. Amer. Chem. Soc. *85*, 337 (1963)
11. White, E. H., Rapaport, E., Hopkins, T. A. and Seliger, H. H., ibid. *91*, 2178 (1968)
12. McElroy, W. D. and Seliger, H. H., Scient. Am. 207, No. 6, 76 (1962)
13. Seliger, H. H. and McElroy, W. D., Science (N. Y.) *138*, 683 (1962)

14. McCapra, F., Chang, Y. C. and Francois, V. P., JCS Chem. Commun. 22 (1968)
15. White, E. H., Steinmetz, M. G., Miano, J. D., Wildes, P. D. and Morlan, R., J. Amer. Chem. Soc. *102,* 3199 (1980)
16. Johnson, F. H. and Shimomura, O., in Methods in Enzymology (DeLuca, M. A., ed.) *57,* 331 (1978)
17. Johnson, F. H. and Shimomura, O., in Methods in Enzymology (DeLuca, M. A., ed.) *57,* 271 (1978)
18. Shimomura, O., Inoue, Y., Johnson, F. H. and Haneda, Y., Comp. Biochem. Physical *B 65,* 435 (1980)
19. McCapra, F. and Chang, Y. C., JCS Chem. Commun. 1011 (1967)
20. Stone, H., Biochem. Biophys. Res. Commun. *31,* 386 (1968)
21. Goto, T., Isobe, M., Coviello, D. A., Kishi, Y. and Inoue, S., Tetrahedron *29,* 2035 (1973)
22. McCapra, F. and Manning, M. J., JCS Chem. Commun., 467 (1973)
23. McCapra, F. and Wrigglesworth, R., ibid. 91 (1969)
24. Hori, K., Wamser, J. E. and Cormier, M. J., ibid. 492 (1973)
25. Goto, T., Pure Appl. Chem. *17,* 4211 (1968)
26. Goto, T. and Fukatsu, H., Tetrahedron Lett., 4299 (1969)
27. Hastings, J. W., in Bioluminescence and Chemiluminescence (De Luca, M. A., ed.) Methods in Enzymology *57,* 125 (1978)
28. Eberhard, A. and Hastings, J. W., Biochem. Biophys. Res. Commun. *47,* 348 (1972)
29. Shannon, P., Presswood, R. B., Spencer, R., Becvar, J. E., Hastings, J. W. and Walsh, C., in Mechanisms of Oxidising Enzymes, (Singer, T. P. and Ondarza, R. F. N., eds.), 69 (1978) Am. Elsevier, N. Y.
30. Bruice, T. C., in Flavins and Flavoproteins (Bray, R. C., Engel, P. C. and Mayhew, S. G., eds.) de Gruyter, New York 1984, p. 45 and references cited
31. Shepherd, P. T. and Bruice, T. C., J. Amer. Chem. Soc. *102,* 7774 (1980) Donovan, V., D. Phil. Thesis, University of Sussex 1980
32. Dixon, B. G. and Schuster, G. B., J. Amer. Chem. Soc. *101,* 3116 (1979)
33. Richardson, W. H., J. Org. Chem. *45,* 303 (1980)
34. Schaap, A. P and Gagnon, S. D., J. Amer. Chem. Soc. *104,* 3504 (1982)
35. Nakamura, H. and Goto, T., Photochem. Photobiol. *30,* 27 (1979)
36. Schaap, A. P., Gagnon, S. D. and Zaklika, K. A., Tetrahedron Lett. *23,* 2943 (1982). Zaklika, K. A., Kissel, T., Thayer, A. L., Burns, P. A. and Schaap, A. P., Photochem. Photobiol. *30,* 35 (1979)
37. McCapra, F. and Perring, K. D., unpublished results
38. Schuster, G. B. and Schmidt, S. P., Adv. Phys. Org. Chem. *18,* 187 (1982)
39. Zaklika, K. A., D. Phil. Thesis, University of Sussex 1976
40. McCapra, F., Beheshti, I., Burford, A., Hann, R. A. and Zaklika, K. A., JCS Chem. Commun., 944 (1977)
41. Shimomura, O. and Johnson, F. H., Biochemistry *7,* 1734, 2574 (1968) Nakatsubo, F., Kishi, Y. and Goto, T., Tetrahedron Lett. 3951 (1969) Shimomura, O., Johnson, F. H. and Kohama, Y., Proc. Nat. Acad. Sci. U.S.A. *69,* 2086 (1972)
42. Ohtsuka, H., Rudie, H. G. and Wamsler, J. E., Biochemistry *15,* 1001 (1976)
43. Dunlop, J. C., Hastings, J. W. and Shimomura, O., FEBS Lett. *135,* 273 (1981)

D. Applications

XIII. Analytical and Other Applications of Chemi- and Bioluminescence

There has been enormous progress in the development of more efficient non – enzymatic chemiluminescent systems – the most outstanding being active oxalic esters and other derivatives (p. 69).

The future of the application of chemiluminescence, however, will probably be not so much as sources of light as such (with the possible exception of special purposes such as safety lights etc.), but in analytical chemistry.

Reasons for this assumption are the very high specifity of certain chemi- and bioluminescent reactions and the commercial availability of very sensitive light measuring devices.

Since such methods rival the sensitivity achieved by radioactive isotopic methods (using principally[125]I) they have an added attraction in their complete safety and freedom from problems of disposal.

Nonetheless, it must be remarked that the patent literature contains an astonishing number of proposals for chemiluminescence light sources.

A short review of these patented claims will be given later in this chapter (see p. 185). As an indication of growing interest three international symposia on analytical applications of chemi- and bioluminescence have recently been held [1–3], and a new journal of bioluminescence and chemiluminescence has been launched [159].

From the enormous number of papers and monographs published in this field a very restricted selection can only be mentioned in this book to give an impression of the potential of chemiluminescence and bioluminescence in analysis.

XIII.A Analytical Applications

It is possible to arrange the analytical applications according to the nature of the chemiluminescent reagents such as luminol, lucigenin, Photinus luciferin etc. It is however probably more useful to discuss them under the various compounds and materials to be analyzed.

Although a complete knowledge of chemiluminescence and bioluminescence reaction mechanisms and the essential factors for the light producing steps is necessary for the optimization of these analytical methods, it is possible in most cases to use them empirically as very sensitive tools in analytical chemistry. Literature published before 1968 has been reviewed [3 a].

XIII.A.1 General and Inorganic Chemistry

XIII.A.1.1 Acid-Base-Titrations

Many chemiluminescent compounds emit light only in neutral or, better in alkaline solution, e.g. luminol and lucigenin. Thus these can be used as neutralization indicators in acid-base titrations. Used in an end-point titration, light can be detected from strongly coloured or turbid or opaque solutions where colorimetric titrations are difficult. For example in the determination of the acidity of milk, of red wine, or mustard [4] or of dark coloured fats and oils [5, 6]. It seems that luminol-fluorescein mixtures are better than luminol alone, because a hemin catalyst is not required here [7]. The system luminol – hemin leads to an irreversible destruction of hemin and is, therefore, not reversible with respect to the catalyst – in contrast to the luminol-fluorescein system [7].

If the solutions to be measured are colourless the precision of the luminol method has been reported to be the same as that of conventional dye indicators [8].

Several combinations in acid-base titrations have been used, for example:

Luminol-hydrogen-peroxide-catalyst [9, 10]; lucigenin-hydrogen peroxide [11, 12]; Luminol-fluorescein-hydrogen peroxide [13] and lucigenin-fluorescein-hydrogen peroxide [14].

XIII.A.1.2 Redox Titrations

It has been pointed out in the preceding chapters that chemiluminescence is intrinsically linked in many cases with oxidation and/or reduction processes. In the luminol chemiluminescence mechanism an early step is the oxidative (dehydrogenative) conversion of the hydrazide to the diazaquinone. This reaction is effected by a large range of one and two electron oxidants, often requiring more than one step. Radicals so produced, or the diazaquinone can react with oxygen or hydrogen peroxide. The catalyst (i.e. analyte) can also react with hydrogen peroxide to produce radicals such as $HO\cdot$ which in turn oxidize luminol. In all cases oxygen or hydrogen peroxide are required for light emission.

XIII.A.1.2.1 Oxidants

One of the first oxidative reactions of luminol was discovered when a luminol solution accidentally reacted with calcium hypochlorite solution [15].

This reaction of hypohalites (hypochlorites and hypobromites) with luminol has since been used as an indicator in oxidative titrations by hypohalite of the following ions: $As^{3(+)} \rightarrow As^{5(+)}$; $Sb^{3(+)} \rightarrow Sb^{5(+)}$; $S_2O_3^{2(-)} \rightarrow S_4O_6^{2(-)}$; $S^{2(-)} \rightarrow S_2^{2(-)}$. Other ions which can be titrated in this way are the rhodanide ($SCN^{(-)}$), the cyanide ($CN^{(-)}$) or the sulfite ($SO_3^{2(-)}$) ions.

Light is emitted with luminol as indicator when the ions are oxidized and the appearance of light indicates that surplus oxidant is present [11, 16]. These titrations are very accurate. Hypochlorite reacts more slowly than hypobromite and elevated temperatures (ca. 80°) are useful in determinations of the former.

Detection limits for hypochlorite itself were found to be 10^{-9} M [17]. It should be pointed out that this detection limit was observed in the absence of peroxides, but that molecular oxygen was required [17].

A similar sensitivity pertains to iodine, but the response to iodine is not linear, with second and third order dependencies being noted [17].

Permanganate ion ($MnO_4^{(-)}$) has a detection limit of 10^{-10} M, and molecular oxygen is not required [17]. Chlorates, chlorites and bromates were found not to interfere, even when present in high concentrations [8].

Of course, hydrogen peroxide can be estimated very precisely by means of the luminol reaction and it is therefore also possible to assay or to follow reactions producing hydrogen peroxide [18–20].

The reaction is catalyzed by appropriate metal ions, with $Co[(NH_3)_4 (NO_3)_2]$ Cl [31] being especially useful, as little as 0.002 μg being effective.

Another chemiluminescent indicator used in this way is lucigenin [11, 21] which also emits light in alkaline solution if there is a surplus of hydrogen peroxide obtained when the oxidation of e.g. As (III) compounds is finished.

Table 12. Determination of Metal ions

Metal	CL-System	References []
Ag	Lucigenin/H_2O_2	17 a)
Bi	Lucigenin/H_2O_2	17 b)
Ce	Luminol/H_2O_2/$Cu^{2(+)}$	17 c)
Co	Luminol/H_2O_2	17 d)
Co		17, 28
Co	Lucigenin/H_2O_2	33
$Cr^{3(+)}$	Luminol/H_2O_2	28
Cu	Luminol/H_2O_2	17 e, f)
Fe	Luminol/H_2O_2/Diethylenetriamine	17 g)
$Fe^{2(+)}$	Luminol/H_2O_2	27
Hg	Luminol/H_2O_2	17 h, i)
$Mn^{2(+)}$	Luminol/H_2O_2 phenanthrolin + citrate as activator	17 k)
Mn	Lucigenin/H_2O_2/amine	31
$Ni^{2(+)}$	Luminol/H_2O_2	17
Os (as OsO_4)	Lucigenin/H_2O_2	33
Pb	Lucigenin/H_2O_2	17 b)
Th	Luminol/H_2O_2	34
Tl	Lucigenin/H_2O_2	17 m)
V	Lucigenin/H_2O_2	17
V	Luminol/H_2O_2 $Co(NH_3)_4 (NO_2)_2$ Cl	31
Zr	Luminol/H_2O_2/$Cu^{2(+)}$	32

The oxidation of siloxene by several strong oxidants [8, 22, 23] has also been suggested as an analytical tool.

XIII.A.2.1.1 Molecular Oxygen; Ozone

In non-aqueous basic solutions such as DMSO/t-butylate, molecular oxygen alone is sufficient to produce luminol chemiluminescence (cf. p. 102). Reactions producing O_2 can be monitored, and a rate of production of 10^{-13} moles per second can be recorded [35]. Ozone also produces luminol chemiluminescence [36], but far better results can be obtained using ozone and rhodamine B.

This fluorescent dyestuff chemiluminesces on treatment with ozone in the liquid phase [37] as well as when adsorbed on silica gel [38]. This latter method has been improved by rendering the surface of the silica-gel hydrophobic [39]. It appears to be the most sensitive method available at present for ozone determination with 1 p.p.b. reported as detectable.

This, by the way, is an example of a chemiluminescence method which can be used very sensitively although the chemical mechanism is not yet known.

XIII.A.1.3 Metal Ions

Table 13 lists the metal ions which have been determined by chemiluminescent methods.

The great sensitivity of luminol to catalytic oxidation by transition metal ions has suggested a variety of analytical procedures. Although many reduceable metal ions cause light emission it is possible to exploit different valency states of the same metal, e.g. chromium since they often exert different catalytic effects on the luminol/hydrogen peroxide reactions, or alternatively one compound in a mixture of catalytically active metal ions can be specifically determined by appropriate complexation procedures [17].

"Classical" metal ions with a positive catalytic effect on the luminol reaction are Fe-(III)-$(CN)_6$-ions or the hemin-Fe-complex.

The latter is the well-known basis for the detection of very small quantities of blood (as in blood-stains) in forensic medicine [24]. Co-(II)-, Cu-(II)-, Cr-(III)-ions have also been determined by the luminol reaction. In addition to highly sensitive photomultipliers, one can also use photographic films for the detection and determination of the chemiluminescence light [25].

Thus, p.p.b. amounts of these metals can be determined. The effect of all these catalysts is thought to be the formation of the hydroxy radical or the superoxide $(O_2^{(\dot{-})})$ – radical anion from H_2O_2.

Direct oxidation of the luminol monoanion is also a possibility, but less likely.

Hydroxyl radicals and/or superoxide radical anions appear to be essential in the luminol reaction in aqueous solution (see p. 101) although not in an aprotic medium.

That the metal catalysis of Cu, Co, Ni and other ions usually pertains to hydrogen peroxide and not to luminol or luminol type hydrazides directly, was demonstrated by the fact that in different hydrazides such as (1)–(3)

XIII (1) XIII (2) XIII (3)

the light yields were observed to follow parallel trends depending on the catalyst and not dependent on the chemical structure of the hydrazide [26].

For optimal use of the luminol chemiluminescence as an analytical procedure, the most effective pH should be provided – in the luminol case, ca. 11.00, depending on the catalyst.

Thus, with Fe(II) and with Cr(III) the pH optimum is 11.00; with Co(II) it is 11.3. Taking $MnO_4^{(-)}$ as oxidant in the luminol reaction, the pH optimum is higher at 12.6. The optimum luminol concentration must be determined for the metal and the oxidant investigated; in general, the concentration range used is 10^{-2} to 10^{-4} M.

For the determination of very low concentrations of metal ions the use of a flow-system [27, 28] is recommended. Figure 20 describes such an apparatus [28].

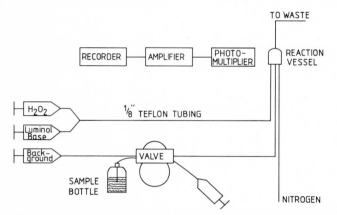

Fig. 20. Flow system (block diagram) for the determination of very low concentrations of metal ions by means of luminol chemiluminescence (after [28]).

Some of these analytical procedures have been applied in the monitoring of biological materials [17] is possible, and upper atmosphere ozone [40] has also been measured.

XIII.A.1.4 Negative (Inhibitory) Metal Catalysis

Use has also been made of a rather unusual observation, that certain metals appear to inhibit luminol chemiluminescence. There is no explanation available as yet for this phenomenon; the suggestion made earlier [158] that these inhibitor metals transform hydrogen peroxide into the non-active $O_2^{2(-)}$ dianion (from which $HO\cdot$-radicals or $O_2^{(\pm)}$ radical anions cannot arise as easily as from O_2H-anion) cannot be correct as in aqueous solution, $O_2^{2(-)}$ dianion is in equilibrium with $HO_2^{(-)}$. Among these metals are vanadium [31], zirconium [32], and thorium [34].

The sensitivity of assays based on this negative effect is $0.04\,\mu g$ to $1.0\,\mu g$.

The procedure is either to add the "inhibitor" solution until the chemiluminescence produced by a luminol/hydrogen peroxide/activator mixture disappears, or to add a complexing agent such as EDTA in excess and to back-titrate, with e.g. a copper-(II)-solution [29].

XIII.A.1.5 Bioluminescent Metal Assays

A series of cations such as $K^{(+)}$, $Ca^{2(+)}$, have been determined by bioluminescence methods. Some assays, especially that of calcium ion determination by aequorin, are very sensitive – even the calcium content of one single cell is within the range of this method. For general application, however, the availability of the respective biological material is decisive [62].

The bioluminescent system of Aequorea is called a photoprotein (aequorin) [64]. The luciferin is bound to the luciferase as a peroxide so that molecular oxygen is not required. The addition of calcium ions "triggers" the luminescence by means of a conformational change of the protein.

The effect of calcium ions is dramatic, the intensity of the resting photoprotein being increased by 6 orders of magnitude. Thus microanalysis of calcium ions is possible with aequorin down to intracellular calcium ion concentrations. Antagonists are $Mg^{2(+)}$ ions, and to a lesser extent $Sr^{2(+)}$ and $Ba^{2(+)}$ ions. Lanthanide ions were found to be still more efficient than calcium ions [65]. Although in vitro calcium analysis can be achieved with highly sensitive $Ca^{2(+)}$ – selective microelectrodes [66], the advantage of the aequorin method appears to be its more rapid response to changes of calcium ion concentration. Thus the main field of application is as an intracellular $Ca^{2(+)}$-indicator in biological systems, rather than in general inorganic analysis. In cellular fluids the problem of aequorin response to other ions is eliminated by the fact that only calcium ions are likely to occur in cells in concentrations necessary to trigger the bioluminescent reaction.

Potassium ions can be determined very precisely by the bioluminescence of the squid Symplectoteuthis [63]. The mechanism of this reaction is not yet fully understood. Sodium ions also stimulate this bioluminescence but the initial intensity is less than 50% of that produced by $K^{(+)}$-ions. The ions $Li^{(+)}$, $Mg^{2(+)}$, and $Ca^{2(+)}$ are far less effective (Table 13).

Table 13. Relative effects of Different Cations on the Initial Light Intensity of Symplecto-theutis Bioluminescence

$K^{(+)}$	100
$Na^{(+)}$	41
$Li^{(+)}$	2.5
$Mg^{(+)}$	4.2
$Ca^{(+)}$	1.6

XIII.A.2 Organic Compounds

In the following section the chemiluminescent assay of some organic compounds which often occur in complex biological mixtures is described.

These reactions are the basis of the empirical use of chemiluminescent analysis of biological substances. There will certainly be improvements by means of the development of modern techniques.

XIII.A.2.1 Amino Acids

As α-amino acids form well-defined copper complexes which catalyze the luminol chemiluminescence, assays have been developed on this basis [41]. This application can be difficult since amino acid copper complexes catalyse the luminol/hydrogen peroxide reaction while amino acids themselves inhibit the luminol/hydrogen peroxide/copper amine reaction [41].

Cysteine can be determined via its inhibitory effect on the luminol/iodine reaction [42].

XIII.A.2.2 Oximes

An inhibitory effect is exerted by oximes on the luminol/hydrogen peroxide reaction [43]. It has been claimed that it mimics certain processes connected with the action of nerve gases on cholinesterase (see next Section) although the connection is tenuous. Oximes are in fact capable of removing the inhibition exerted on cholinesterase by nerve gases [44] and it has been suggested that chemiluminescence can be used to monitor this.

XIII.A.2.3 Organophosphorus Compounds

It has been suggested that nerve gases such as Sarin (4) can be determined with luminol/sodium perborate/trisodium phosphate with a sensitivity down to 0.5 µg [45].

The basis of this analysis is said to be the formation of a peroxy phosphate compound (the Schoenemann reaction [49]). Oxidation of an amine such as 3,3′-dimethoxy benzidine or indole in a colorimetric reaction, or in the present case luminol with the emission of light then follows. Lucigenin does not work in this assay although it has been claimed as useful in the analysis of the nerve gas Tabun (5) [46]. Related insecticides such as Isopestox behave similarly [46–48]. Sensitivity can be improved by addition of chloride ion [52], enhancing the light yield, or by suppression of background chemiluminescence by complexation of trace transition metals by EDTA.

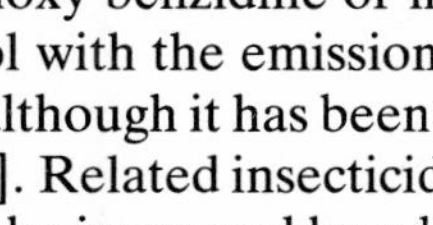

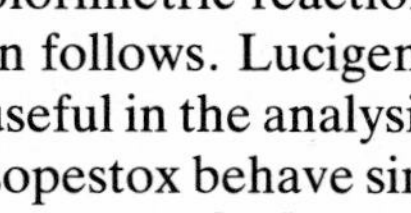

173

Less toxic (and less reactive) alkyl phosphate esters such as trimethyl-, tributyl- and tricresylphosphate do not interfere in amounts below 20 µg.

XIII.A.2.4 Phenols, Naphthols

These phenolic compounds were determined by means of their inhibitory effect on the luminol/H_2O_2 chemiluminescence catalyzed by copper/ammonia [51] or by cobalt-(II)-ions [52].

The detection limit was reported to be 400 µg with phenols, 20 µg with certain naphthols.

XIII.A.2.5 Autoxidations

On p. 19 the chemiluminescence of autoxidation reactions was described. Although it is mostly very weak, often detectable only by means of energy transfer to suitable efficient fluorescers, it has been fairly frequently used as a sensitive analytical tool. Sufficiently sensitive light measuring equipment of the single photon counting type is required.

The very low efficiencies of the reactions involved make mechanistic investigation difficult, but recently the light emission has been used for several very practical purposes. These include investigation of the oxidative degradation of long-chain aliphatic residues, such as those in fatty acids, fats, oils and polymers.

By extension the efficacy of anti-oxidants and stabilizers which interrupt the radical oxidation chain responsible for chemiluminescence can be explored. Similarly the efficiency of radical chain initiators can be examined.

These applications allow the measurement of the expected life time of lipid-rich foods and of structural polymers such as polyethylene. Such is the sensitivity of modern light detecting devices that the very low concentration of O_2 involved can also be determined.

XIII.A.2.5.1 Oxidative Degradation of Molecules Containing Long-Chain Aliphatic Residues

This is important in food analysis and technology, as affecting also the stability of plastics under every-day conditions, i.e. in the presence of air and light. The chemiluminescence observed is extremely weak (yields of 10^{-15} einstein/mole^{-1} or less). Such light emission is very probably caused by peroxy radical recombinations (see p. 23). Measurements of the spectra of such ultraweak chemiluminescence have been achieved [53–55].

The evaluation of the degree of deterioration in food has been developed recently [56]. Almost all the loss of quality (changes in flavour particularly) is the result of oxidation. This is exemplified by the effect of molecular oxygen on the chemiluminescence intensity of a commercial salad oil at 45° as seen in Fig. 21.

Using this technique, other foodstuffs (soy bean oil, potato frying oil, Chinese noodles etc.) can be controlled with respect to their oxidative stability and the effect of certain additives.

These applications have been reviewed [57]. Detailed investigations, mostly qualitative, on autoxidation chemiluminescence have been performed on methyl

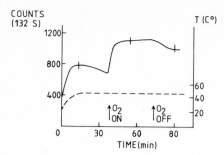

Fig. 21. Changes of emission intensity of commercial salad oil with or without bubbling oxygen at 45 °C [56].

oleate [58], t-butyl oleate [59], sodium linoleate [60] and on linoleic and docosahexanoic acid films [61].

XIII.A.2.5.2 Efficiency of Radical Chain Inhibitors and Initiators

Under steady state conditions the formation of radicals starting an autoxidation chain process is equal to their disappearence in the chain terminating step [67].

The chemiluminescence intensity of an appropriate autoxidation system is a function of the steady state radical concentration. If radical scavengers diminish that concentration, their efficiency can be directly measured by means of the steady state chemiluminescence [68].

An example of the process is seen in Figure 22.

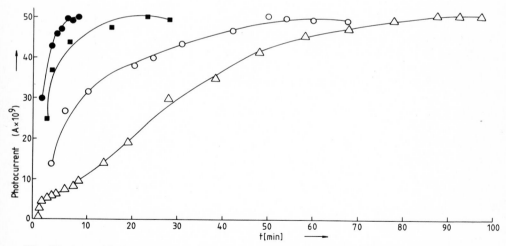

Fig. 22. Autoxidation Chemiluminescence as a Function of Radical Scavengers, in Polypropylene (after [57]).

●:   without antioxidant (inhibitor)
■:   with the antioxidant dilauryl-3,3'-thiodipropionate (0.01%) (A)
○:   with the antioxidant 6,6'- di-t.-butyl -4,4'-thiodi(o-cresole) (B) (0.01%)
△:   with the antioxidant mixture A + B, 0.01% each (synergistic effect)

By contrast, autoxidation chemiluminescence is enhanced by initiators whose efficiency, therefore, can also be assayed by chemiluminescence methods as shown in Fig. 23.

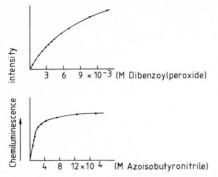

Fig. 23. Effect of Radical Chain Initiators on Autoxidation Chemiluminescence [68]. In Fig. 23:1) Initiator: Dibenzoylperoxide (in Ethylbenzene), in Fig. 23:2) Initiator: Azo-isobutyronitrile in Ethylbenzene-Acetic Acid

XIII.A.2.5.3 Determination of Low Concentrations of Molecular Oxygen; the "Oxygen Drop"

Oxygen is necessary in autoxidation chemiluminescence for the formation of peroxy radicals (ROO·) the recombination of which is the cause of the chemiluminescence (see p. 23). On the other hand, molecular (triplet) oxygen is a strong chemiluminescence quencher. As a result the intensity vs.time curve rises as an approximately linear function whereas the oxygen concentration simultaneously drops. The more the oxygen originally present (in a closed system) is consumed the more its inhibitory effect on autoxidation chemiluminescence also diminishes. When all the oxygen is consumed the chemiluminescence ceases abruptly [69].

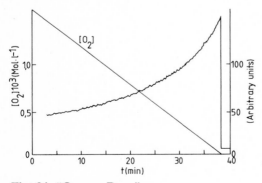

Fig. 24. "Oxygen Drop"
Intensity vs. time plot of the chemiluminescence of ethylbenzene autoxidation in benzene, at 40 °C. Starter: dicyclohexyl peroxycarbonate (5.2×10^{-2} M). After Vasil'ev and Rusina [69].

176

XIII.A.2.5.4 Other Methods of Oxygen Determination

Tetrakis-dialkylamino ethenes, e.g. the dimethylamino derivative TMAE are oxidized to 2 moles of tetra-alkyl urea (p. 119).

This is accompanied by chemiluminescence, the emission of which matches the fluorescence of TMAE or that of the respective reactant [70–72]. This reaction has been used for oxygen determination in the p.p.m.-range [73].

Luminescent bacteria are extraordinarily sensitive to oxygen and have been used as sensors [160].

XIII.A.2.6 Glucose

Glucose is determined indirectly: on treatment with glucose oxidase it is transformed into gluconic acid (via gluconolactone) and hydrogen peroxide:

$$\text{Glucose} + O_2 + H_2O \xrightarrow{\text{glucose oxidase}} \text{gluconic acid} + H_2O_2$$

The last can be assayed either in the luminol reaction [74, 75], or by an appropriate activated oxalate, e.g. 2,4,6-trichlorophenyl oxalate, with perylene as fluorescer [76]. The detection limit of glucose in the latter system is 7×10^{-8} M.

Another more esoteric possibility is the use of earthworm (Diplocardia) bioluminescence for hydrogen peroxide determination [62].

Whereas this very sensitive method (sensitivity ca. 10 picomoles) is dependent on the availability of the bioluminescent system, chemical methods present no such problem. One of the most effective systems is that using acridinium phenyl esters (p. 115).

Extremely good precision is obtained (coefficient of variation 0.6%) and an excellent correlation with blood glucose measured directly in serum or in whole blood [77].

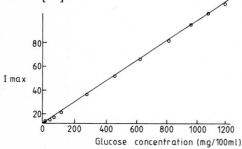

Fig. 25. Variation of I_{max} as a function of glucose concentration, using acridinium p-acetyl phenyl carboxylate as detector of H_2O_2 produced by glucose oxidase at pH 8.

XIII.A.2.7 Alcohols and Aldehydes

When alcohols react with oxygen on the surface of a solid alkalimetal hydroxide, a weak chemiluminescence is observed which can be quantitively determined by photon counting techniques [78]. This phenomenon is of interest in the search for a cause of the background chemiluminescence of basic solutions of biological materials.

The corresponding aldehydes respond faster to the presence of alkali metal hydroxides than do the alcohols.

The alkali leads to aldol condensations and polymerizations [79]. The severe conditions and the complexity of the reaction make mechanistic interpretations very problematical. It is not clear what value this approach has as an analytical technique since interference by other easily oxidized materials will be troublesome.

XIII.A.3 Biological Systems

Of all the applications of bioluminescence, that of the firefly reaction is by far the most important.

According to the mechanism described on p. 152 it is dependent on ATP, therefore, on one of the most universal compounds in living matter. ATP is linked with so many essential biochemical reactions that a sensitive method for its assay is of paramount importance, and applications of ATP dependent bioluminescence are found in clinical chemistry, marine biology, soil science etc.

Concentrations of ATP as low as 10^{-15} mol (femtomoles) can be determined – quantities corresponding to the ATP content of single cells [80]. In this section, only a few of the examples in this enormous field of biochemical analysis can be mentioned. More information can be obtained in a monograph [81] in which such biochemical applications are described in detail.

XIII.A.3.1 ATP Assay with Firefly Bioluminescence

The first analytical use of firefly luciferase was described more than 30 years ago [82].

As ATP is a "universal" biochemical reagent, in so far as it is the product of, or as it is involved in, a multitude of biochemical reactions, firefly bioluminescence can be used as an indicator of a variety of coupled reactions, and the pathways of many biological processes inferred indirectly.

Other triphosphates, such as ITP, GTP, cannot replace ATP in the reaction, and the same is true of the corresponding diphosphates such as ADP and IDP [84].

For practical use it is very important that ATP not only produces one short burst of light, the intensity of which is proportional to ATP concentration, but that conditions are chosen which produce a constant light signal of some duration [84].

These conditions were first described in 1953 [85]. Some years later (1975), they were rediscovered [86], and their analytical significance pointed out [87]. An explanation for the kinetics leading to this constant signal has been proposed; on the basis of these ideas, a light signal of about 20 min duration can be obtained.

The most significant feature of this mechanism is that in a complex of luciferase and oxidated luciferin (L), the enzyme has undergone a conformational change which inhibits the dissociation of the enzyme-L complex [88].

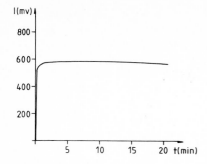

Fig. 26. Constant Signal of Firefly-ATP-Assay [88].

Experimental conditions for obtaining such a constant signal are a large excess of luciferase, compared with the amount of product formed during the reaction and avoidance of reduction of the reaction rate by inhibitors such as pyrophosphate (see also [89]).

XIII.A.3.1.1 Scope of the ATP Assay

The scope of the ATP assay is wide and it has been used to monitor the cellular level of ATP, cell lysis, fermentation rates in foods and beverages, microbial activity [98] and ATP synthesis in chloroplasts [99]. Detection and counting of micro organisms of various kinds in for example fresh [94] and waste [95] water is a generally useful application. In view of the ubiquity of ATP it is not surprising to find a growing list [1, 2, 81, 83] of uses for the technique.

XIII.A.3.1.2 Creatine Kinase and Pyruvate Kinase

Kinases which generate ATP can be assayed specifically, and creatine kinase, in view of its importance in the diagnosis of heart disease (myocardial infarction) has been much investigated and refined [103, 104]. In principle any kinase generating or using ATP can be assayed and the method is among the most rapid and sensitive available. The reactions catalyzed by two kinases are shown:

$$\text{creatine phosphate} + \text{ADP} \xrightleftharpoons{\text{creatine kinase}} \text{creatine} + \text{ATP}$$

$$\text{phosphoenol pyruvate} + \text{ADP} \xrightleftharpoons{\text{pyruvate kinase}} \text{pyruvate} + \text{ATP}$$

XIII.A.3.2 Analytical Applications of Bacterial Bioluminescence

An essential feature of bacterial bioluminescence is the requirement for long chain aldehyde (p. 156). Some bacteria, or certain mutants thereof, can use long-chain fatty acids, especially myristic acid [107] as precursors of the aldehydes. Thus a dark mutant of B. harveyi emits light when myristic acid is added. The light yield is proportional to the quantity of the acid added, down to 10 picomoles per ml.

XIII.A.3.2.1 Activity of Lipases

The activity of lipases can thus be determined by their action on glycerides, containing myristic acid acyl groups, and bioluminescent assay of the liberated myristic acid with B. harveyi. Phospholipases A_2 and C have also been investigated in this manner [108].

XIII.A.3.2.2 Mutagenic Activity

The mutagenic activity of chemical substances can be assessed by means of the Ames test [109]. In analogy to this test which is based on the reversion of certain changes in the phenotype of mutants caused by the mutagen, one can use the reversion of a "dark" variant of Photobacterium leiognathi to the luminous one by mutagens, such as mitomycin C. This can be detected in nanogram amounts [110].

XIII.A.3.2.3 NAD(P)H-Assay

Bacterial luciferase can be used in the assay of NAD(P)H by a coupling technique. In the first step flavin mononucleotide (FMN) is reduced by NAD (P)H and NADH/FMN-oxidoreductase. $FMNH_2$ then reacts with oxygen, a long chain aldehyde, and bacterial luciferase to give FMN, carboxylic acid, water and light (as described on p. 156) [111].

XIII.A.4 Phagocytosis

There have been many reports over many years of luminescence arising from ordinary cells, in no way connected with bioluminescence as such. The early reports were of very weak radiation, and it is now clear that a large variety of organisms emit light during normal growth [161]. Cilento and his collaborators have shown that many oxidative enzymic reactions are capable of generating light [111 a] and have speculated on its possible role. Mechanisms elucidated in this work may help explain the light emitted during phagocytosis. At present, however, this light is of most interest as an investigative tool in the study of the process itself.

When granulocytic white blood cells engulf opsonized* bacteria, chemiluminescence is observed [112].

The chemiluminescence can be enhanced by luminol [113, 113 a, 114] or by lucigenin [115].

The complicated mechanism cannot be dealt with in detail here, but the final event appears to be the oxidation of intracellular NADPH by oxygen and NADPH-oxidase, to yield NADP and hydrogen peroxide [116]. The chemiluminescence yield observed correlates with the microbiocidal action of the leucocytes [116].

In the absence of luminol or lucigenin, electronically excited carbonyl groups are thought to be the emitters of "native" phagocytosis chemiluminescence [117], but the enhanced chemiluminescence obtained in the presence of luminol or lucigenin, stems from 3-aminophthalate and N-methylacridone, respectively, produced very probably by the action of $O_2^{(-)}$ radical anion.

---

* Opsonization: Complexation of bacteria by specific immunoproteins (Opsonins) of the blood serum.

It has been suggested but now seems rather improbable, that singlet oxygen may be involved in this phagocytosis chemiluminescence [118]. This conclusion is supported experimentally by means of the most reliable method available at present for the "chemical" detection of singlet oxygen: its unambiguous reaction with cholesterol (p. 13).

XIII.A.5 Chemiluminescent Immunoassays

The high specifity of antigen-antibody reactions, and the sensitivity of these reactions, combined with the equally high sensitivity of chemi- and bioluminescence reactions, have led to the development of methods of chemiluminescent immunoassays, as an alternative to the commonly used radioimmunoassay [119].

In the latter method, either the antigen or the antibody is labeled with a radioactive isotope – e.g. ^{125}I.

Radioimmunoassay is at present widely used, particularly in routine methods in clinical laboratories, but it has some serious problems, concerning stability of the labeled material [120], and the disposal of the radioactive wastes.

Two general methods are available at present for chemiluminescent immunoassay, in which the *antigen* is labeled by a component of a chemi- or bioluminescent system:

I. The antigen is combined with a chemiluminescent group by covalent bonds. This "conjugate" is reacted with the antibody, together with "free" antigen, in the competitive formation of the antigen-antibody complex (see later for an example). After separation of the bound and free antigen in the usual way, light emission then replaces radioactivity as the means of quantification. By treatment of the antibody-"conjugate" complex with an appropriate oxidizing reagent, e.g. hydrogen/microperoxidase, chemiluminescence is produced whose quantum yield is proportional to the antibody concentration.

II. The antigen is coupled with an enzyme, which catalyzes the chemiluminescence reaction of an appropriate compound plus an oxidant.
   Horse radish peroxidase and firefly luciferase have been used as such enzymes. The competitive antigen-antibody complex is formed as in I. Chemiluminescence occurs when this complex is treated with a chemiluminescent compound and an oxidant. In the case of peroxidase linked to the antigen, luminol and related compounds have been used. In the case of luciferase, ATP, firefly luciferin, magnesium ions, and oxygen produce the light as usual (p. 152).

Examples of method I are the immunoassay for biotin [121], for serum thyroxine [122], and for hepatitis B surface antigen [123]. Conjugates used were the isoluminol derivatives (6), (7), and (8)

XIII (6)

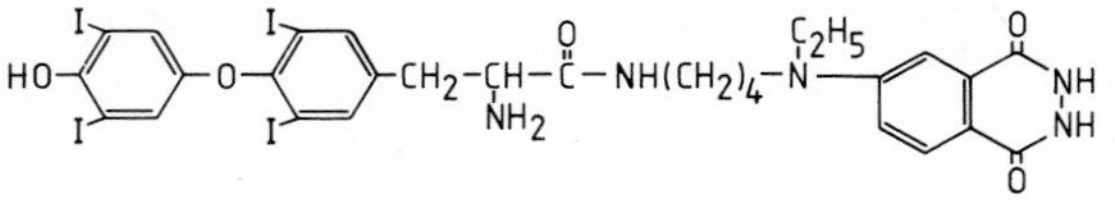

$$\underline{XIII}(7)$$

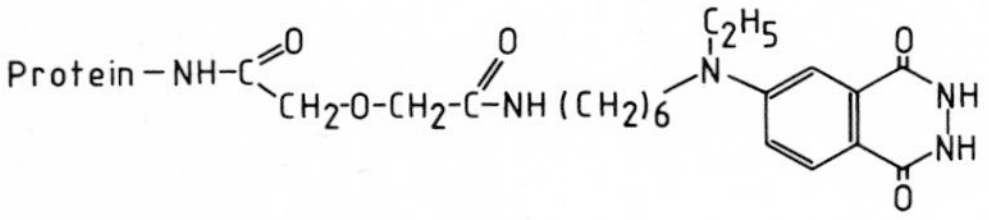

$$\underline{XIII}(8)$$

After formation of the antigen-antibody complex, and removal of excess of reagents, light was produced by reacting the complex with microperoxidase*/hydrogen peroxide, or lactoperoxidase/hydrogen peroxide (the latter oxidative system was used only in the biotin chemiluminescent immunoassay).

In the chemiluminescent immunoassay of steriods such as progesterone or 17 β-estradiol, either isoluminol [125, 125 a, 126] or acridinium esters have been used. It is claimed that the latter offer some advantages over luminol or isoluminol, insofar as they require hydrogen peroxide and low base concentrations only, without an additional catalyst [127, 127 a].

A conjugate of this type is (7) [127] or (8) [127 a]:

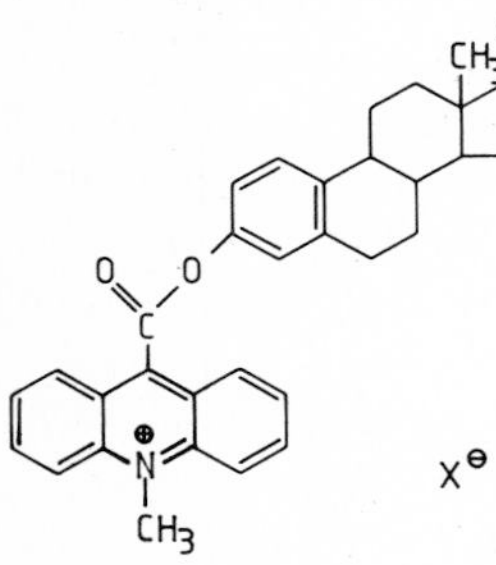

$$\underline{XIII}(7)$$

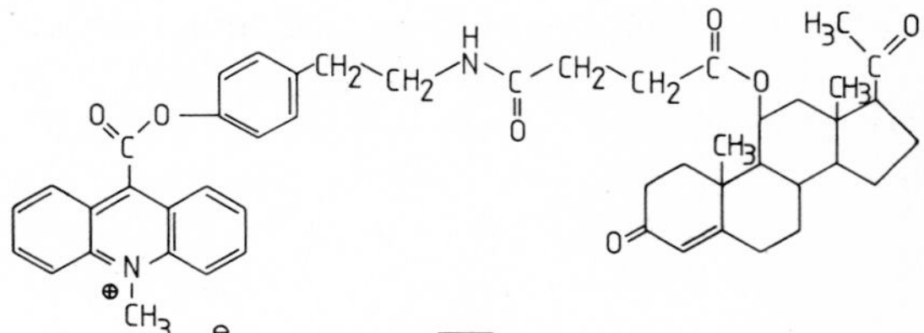

$$\underline{XIII}(8)$$

The chemiluminescent group of (7) is linked directly to the steroid structure. Direct coupling has also been achieved between the acridinium structure and sheep (anti-AFP) antibody by means of a carbodiimide reaction [128].

Plasma cortisol and pregnanediol have been determined by chemiluminescence immunoassay [90, 91, 125 a]; (9) was synthesized as the conjugate.

---

* Microperoxidase/H_2O_2 was found to be extremely effective in the whole pH-range of 8.6–13 [124].

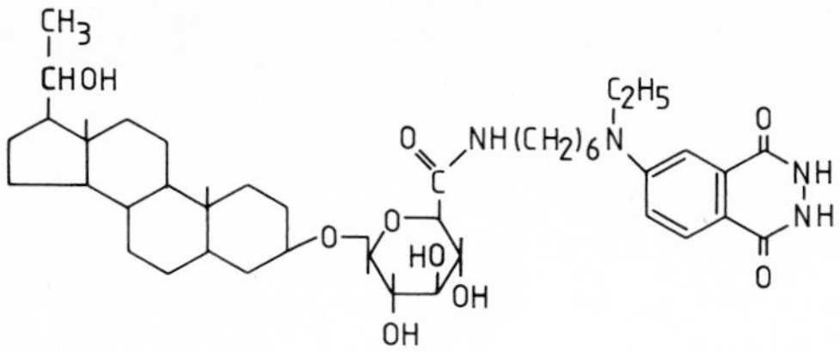

XIII (9)

Enzyme immunoassays for digoxin, thyroxine and insulin have been performed with horse radish peroxidase as a label. The antigen-antibody complex was treated with luminol or other related hydrazides, the most efficient being 7-dimethylamino naphthalene-1,2-dicarboxylic acid hydrazide (92) (cf. p. 80) [129].

Immunoassays have been developed by labelling the *antibody*. Acridinium esters can be linked to the immunoglobulin by use of an N-hydroxysuccinimidyl ester grouping. Although the conjugate is not completely stable (the phenyl ester hydrolyses slowly in the buffer) this is one of the most promising of the new techniques. Sandwich assays for important serum constituents such as α-fetoprotein [128 a], thyroid stimulating hormone (TSH) and antibodies to hepatitis B are all more sensitive, at least as accurate and more convenient than existing methods.

An interesting application of the chemiluminescent energy transfer reactions described on p. 85 is the use of an antibody labelled with a fluorescent acceptor (such as fluorescein) and an antigen attached to a chemiluminescent compound like luminol [129 a]. The principle behind this idea is that the unbound antigen will emit the blue light characteristic of luminol, and the bound fraction will, because of the proximity of donor and acceptor, emit the fluorescence of fluorescein. The energy transfer occurs by the resonance mechanism, and only those species within very short distances will undergo energy transfer. By using two different filters in the detecting system, the proportions of bound and unbound antigen can be detected. This means that no separation of bound and unbound fractions are required i. e. a homogeneous assay. If chemiluminescence immunoassays are to be used as sensitive probes of activity in living cells, this is a necessary development. The method has been applied with great success to the analysis of the secondary messenger c-AMP.

XIII.A.5.1 Drug Assay

Clinically important drugs, such as methotrexate 4-amino-10-methyl folic acid MTX – which has considerable significance as an antileukaemic agent –, can be assayed using chemiluminescence immunoassay methods, coupling the drug with firefly luciferase.

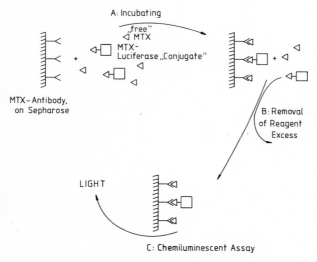

MTX

The methotrexate assay [130] is described in some detail, as it can serve as an example of the chemiluminescence immunoassay method in general [131].

Methotrexate (MTX) is coupled to firefly luciferase using a water-soluble carbodiimide.

It was found that 2 moles of MTX per mole of luciferase had reacted on average. The coupling product retained 70% of the enzymatic activity of the luciferase.

This coupling product was reacted, in competition with free MTX, with MTX antibody, immobilized on a carrier (in this case, sepharose was used). After formation of the complexes, and removal of surplus reagents, light production and measuring is performed. Fig. 27 describes the scheme of this chemiluminescent immunoassay:

A: Incubating

"free" MTX

MTX-Luciferase "Conjugate"

MTX-Antibody, on Sepharose

B: Removal of Reagent Excess

LIGHT

C: Chemiluminescent Assay

Fig. 27. Methotrexate Chemiluminescence Immunoassay [131].

The competitive binding curve of free methotrexate and luciferase-methotrexate is seen in Fig. 28.

The curve in Fig. 28 was obtained by incubating a constant quantity of MTX-luciferase conjugate with increasing amounts of free MTX: there is a linear relation between light yield and free MTX from 0 to 10 pmoles.

In an analogous way, the competitive binding of antigen labeled with either a

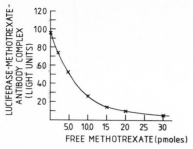

Fig. 28. Competitive binding curve of free methotrexate with luciferase-methotrexate [131].

chemiluminescent compound, or peroxidase or luciferase with free antigen serves as basis for the determination of the respective antigen.

XIII.B Other Applications

Numerous applications (other than the analytical uses described in chapter XIII A) of chemiluminescent systems have been proposed. This is evident from the very large number of patent applications, only a few of which can be mentioned.

The application of chemiluminescence as a light source is most successful when a particular requirement is met such as freedom from heat or ignition. Such uses are as an emergency light, for signalling and marking, in special display devices, and in indicator systems involving oxidants.

Obviously the most important requirements for a chemiluminescent system used in one of the above applications are as high a quantum yield as possible, combined with a high light intensity. The reagents must have a long shelf life. The materials must be readily synthesized and cheap.

XIII.B.1 Emergency Lighting

The most common chemiluminescent emergency lights – which are also commercially available – are the well known Cyalume(R)Lightsticks [132] based on *peroxy oxalate chemiluminescence* (p. 69). Certain trifluoroacetates, e. g. $(CH_3)_4 N^{(+)} CF_3CO_2^{(-)}$, improve the quantum yield of the system by ca. 50% [133, 134]. Another emergency light suggested for rescue operations is produced by the spontaneous autoxidation of *tetrakis-dialkylamino-ethylenes* in air (p. 119) [135] Processes for their production [136], optimization of their chemiluminescence by removal of inhibition by use of inorganic salts [137], and emission of orange light by energy-transfer (e. g. to Rhodamine B) [138] have been patented.

185

XIII.B.2 Signalling and Marking

This can be achieved by chemiluminescent smokes [139, 140], aerosols and sprays [141–143], or emulsions in water [144] – all of them containing *tetrakis-dialkylamino ethylenes.*

The *peroxyoxalate* chemiluminescence is used here, too: "warning capsules against infiltrating troops" were developed containing an active oxalate, a fluorescer, and – as a hydrogen peroxide source, producing H_2O_2 on contact with air – an anthrahydroquinone derivative (258).

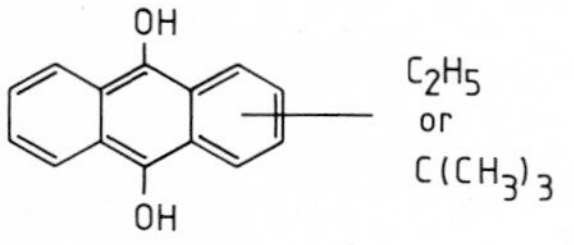

$\underline{XIII}$ (10)

[145]. Chemiluminescent smokes are also produced by the reaction of trialkyl aluminium compounds, complexed with an appropriate ether, with air, in the presence of acridine derivatives [146], indoles [147], and other compounds [148–150].

XIII.B.3 Special Display Devices

Electrogenerated radical ion chemiluminescence, produced on special coated electrodes, is claimed to be suitable as display elements [151], in indicator screens, as lasers with electrochemical pumping [152], or in computer and automation technology [153].

Electrochemiluminescent coated electrodes have also been described [154].

XIII.B.4 As Indicator Systems in Processes Involving Oxidants

Detergents containing peroxides, such as perborates, will produce chemiluminescence with luminol or other appropriate compounds only when the cleaning process or a disinfection procedure is complete, i. e. when there is a surplus of oxidant [155].

XIII.C References

1. Schram, E., and Stanley, P., (Eds.), Proceedings of the International Symposium on Analytical Applications of Bioluminescence and Chemiluminescence, State Printing and Publishing Inc., Westlake Village Calif. 1979
2. De Luca, M. A., and McElroy, W. D., (Eds), Bioluminescence and Chemiluminescence – Basic Chemistry and Analytical Applications Academic Press, New York 1981

3. Kricka, L. J., Stanley, P. E., Thorpe, G. H. G. and Whitehead, T. P., (Eds.) Analytical Applications of Bioluminescence and Chemiluminescence, Academic Press, New York 1984
3a. Gundermann, K.-D., Chemilumineszenz organischer Verbindungen, p. 159, Springer-Verlag, Berlin 1968
4. Erdey, L., Ind. Eng. Chem. *33*, 459 (1957)
5. Yarovenko, E. Ya. and Kosheleva, K. N., Zavod. Lab. *27*, 407 (1961)
6. Kovats, M. T. and Paal, Z., Magy. Kem. Lapja *21*, 489 (1966)
7. Erdey, L., Pickering, W. F. and Wilson, C. L., Talanta *9*, 371 (1962)
8. Isaacson, U. and Wettermark, G., Analyt. Chim. Acta *68*, 339 (1974)
9. Kenny, F. and Kurtz, R. B., Anal. Chem. *23*, 339 (1951)
10. Kenny, F. and Kurtz, R. B., ibid. 1218 (1952)
11. Erdey, L., Ind. Chem. *33*, 459, 523, 575, (1957)
12. Erdey, L., Acta Chem. Acad. Sci. Hung. *3*, 81 (1953)
13. Kenny, F. and Kurtz, R. B., Anal. Chem. *25*, 1550 (1953)
14. Kenny, F., Kurtz, R. B., Beck, I. and Lukosevicius, I., ibid. *29*, 543 (1957)
15. Albrecht, H. O., Z. physik. Chem. *136*, 321 (1928)
16. Erdey, L. and Buzàs, I., Acta Chim. Acad. Sci. Hung. *6*, 93, 115, 123 (1955)
17. Seitz, W. R. and Hercules, D. M., in: Cormier, M. J., Hercules, D. M. and Lee, J., (Eds.) Chemiluminescence and Bioluminescence, p. 427, Plenum Press, New York 1973.
17a. Babko, A. K., Terletskaya, A. V. and Dubovenko, L. I., J. Anal. Chem. USSR *23*, 809 (1968)
17b. Dubovenko, L. I. and Khotinets, F. Ya., Zh. Anal. Khim. *26*, 784 (1971)
17c. Dubovenko, L. I. and Huu, C. T., Ukr. Khim. Zh. *35*, 637 (1969)
17d. Babko, A. K. and Lukovskaya, N. M., Zavod. Lab. *29*, 404 (1963)
17e. Babko, A. K. and Lukovskaya, N. M., J. Anal. Chem. USSR *17*, 47 (1962)
17f. Dubovenko, L. I. and Pihpenko, L. A., Visn. Küv. Univ. Ser.Khim., *11*, 75 (1970)
17g. Kalinichenko, I. E. and Grishchenko, O. M., Ukr. Khim. Zh. *36*, 610 (1970)
17h. Dubovenko, L. I. and Tovmasyan, A. P., ibid. *37*, 943 (1971)
17i. Dubovenko, L. I. and Guz, L. D., ibid., *36*, 1264 (1970)
17k. Dubovenko, L. I., and Tovmasyan, A.P., ibid., *37*, 845 (1971)
18. Ojima, H. and Iwaki, R., Nippon Kagaku Zasshi *78*, 1632 (1957)
19. Langenbeck, W. and Ruge, U., Ber. dt. chem. Ges. *70*, 367 (1937)
20. Weber, K. and Flesh, D., Nature *191*, 177 (1961); and previous papers
21. Erdey, L., and Buzàs, I., Acta. Chim. Acad. Sci. Hung. *6*, 77 (1955)
22. Kenny, F. and Kurtz, R. B., Anal. Chem. *22*, 693 (1950)
23. Erdey, L., Buzàs, I. and Pólos, L., Z. Anal. Chem. *169*, 187 (1959)
24. Steigmann, A., Chem. Ind. *60*, 889 (1941)
25. Babko, A. K. and Kalinichenko, I. E., Ukr. Khim. Zhurn. *31*, 1316 (1965)
26. Gundermann, K.-D. and Hercules, D. M., unpublished results
27. Seitz, W. R. and Hercules, D. M., Anal. Chem. *44*, 2143 (1972)
28. Seitz, W. R., Suydam, W. W. and Hercules, D. M., ibid., *44*, 957 (1972)
29. Erdey, L. and Buzàs, I., Anal. Chim. Acta *22*, 524 (1960)
30. Erdey, L., Weber, O. and Buzàs, I., Talanta *17*, 1221 (1970)
31. Babko, A. K. and Lukovskaya, N. M., J. anal. Chem. USSR *20*, 1153 (1965)
32. Babko, A. K., Dovenko, L. I. and Mikhailova, L. S., Metody Anal. Khim. Reakt. Prep. No. *13*, 139 (1969)
33. Bognar, J. and Sipos, L., Mikrochim. Ichnoanal. Acta *5–6*, 1066 (1963)
34. Dubovenko, L. I. and Huu, C. T., Ukr. Khim. Zhurn. *35*, 957 (1957)
35. Buri, A. and Mayzerall, D., Biochim. Biophys. Acta *153*, 614 (1968)

36. Regener, V. H., J. Geophys. Res. *65*, 3975 (1960)
37. Bersis, D. and Vassiliou, E., Analyst *91*, 499 (1966)
38. Regener, V. H., J. Geophys. Res. *69*, 3795 (1964)
39. Hodgeson, J. A., Krost, K. J., O'Keefe, A. E. and Stevens, R. K., Anal. Chem. *42*, 1795 (1970)
40. Randhawa, J. S., Nature *213*, 53 (1967)
41. Ponomarenko, A. A. and Amelina, L. M., J. Gen. Chem. USSR, *35*, 750, 2252 (1965)
42. Lukovskaya, N. M. and Markova, L. V., J. Anal. Chem. USSR *24*, 1512 (1964)
43. Weber, K., Matkovic, J. and Fles, D., Nature *91*, 177 (1961)
44. Matkovic, J., and Weber, K., Arch. Toxikol. *21*, 355 (1966)
45. Goldenson, J., Anal. Chem. *26*, 877 (1956)
46. Weber, K. and Matkovic, J., Arch. Toxikol. *21*, 38 (1965)
47. Weber, K., Huic, L. and Mrazovic, M., Arh. Hig. Rada *9*, 325 (1958)
48. Weber, K., Matkovic, J. and Busljeta, M., Acta Pharm. Jugosl. *19*, 47 (1969)
49. Fritsche, U., Anal. Chim. Acta *118*, 179 (1980)
50. Weber, K. and Matkovic, J., Arh. Hig. Rada *15*, 151 (1964)
51. Ponomarenko, A. A. and Popov, B. I., J. Anal. Chem. USSR *19*, 1300 (1964)
52. Lukovskaya, N. M. and Gerasimenko, M. I., ibid. *26*, 1462 (1971)
53. Inaba, H., Kaneko, N. and Ichimura, T., J. Quantum Electronics QE-*5*, 320 (1969)
54. Inaba, H., Shimizu, Y. and Tsuji, Y., Jap. J. Appl. Phys. *14*, Suppl. 14-1, 13 (1975)
55. Shimizu, Y., Inaba, H., Kumaki, K., Mizuno, K., Hata, S. and Tomioka, S., IEEE Trans. Instrum. Meas. IM-22, 153 (1973)
56. Usuki, R., Kaneda, T., Yamagishi, A., Takyu, C. and Inaba, H., 6th Intern. Congress of Food Sci. and Technol., Kyoto 1978, see J. Food Sci. *44*, 1573 (1979)
57. Mendenhall, G. D., Angew. Chem., Int. Ed. Engl. *16*, 225 (1977)
58. Vassil'ev R. F. and Vichutinskii, A. A., Nature *162*, 1276 (1962)
59. Höfert, M., Angew. Chem. *76*, 826 (1964)
60. Mizuno, K., Hata, S., Kumaki, K., Inaba, H. and Shimizu, Y., Proc. 94th Ann. Meeting, Pharm. Soc. Jap. 1974
61. Slawson, V. and Adamson, A. W., Lipids *11*, 472 (1976)
62. Wampler, J. E., in Lit. 2), p. 249
63. Leisman, G. and Tsuji, F. I., in Lit. 2), p. 709
64. Blinks, J. R., in Lit. 2), p. 243; Hallett M. B. and Campbell, A. K., in: Clinical and Biochemical Luminescence, Kricka, L. J. and Carter, T. N. Eds., M. Dekker, N. Y. 1982, p. 89
65. Blinks, J. R., Allen, D. G., Prendergast, F. G. and Harrer, G. C., Life Sciences *22*, 1237 (1978)
66. Ammann, D. P., Meir, P. C. and Simon, W., in Detection and Measurement of Free Calcium Ions in Cells p. 73, (Ashley, C. C. and Campbell, A. K., Eds.: Elsevier, Amsterdam 1979)
67. Papisova, V. I., Shlyapintokh, V. Ya. and Vassil'ey, R. F., Usp. Chim. *34*, 599 (1965)
68. Shlyapintokh, V. Ya., Karpukhin, O. N., Postnikov, L. M., Tsepalo, V. F., Vichutinskii, A. A. and Zakharov, I. V., Chemiluminescence Technique in Chemical Reactions, Consultants Bureau, New York 1968
69. Vasil'ev, R. F. and Rusina, I., Doklad, Akad. SSSR *153*, 1101 (1963) see also: Worthy, W., Chem. Engl. News. p. 30, Nov. 24, 1975
70. Urry, W. H. and Sheeto, J., Photochem. Photobiol. *4*, 1067 (1965)
71. Paris, J. P., ibid. *4*, 1059 (1965)
72. Fletcher, A. N. and Heller, C. A., J. Phys. Chem. *71*, 1507 (1967)

73. Freeman, T. M. and Seitz, W. R., in Lit. 2), p. 347
74. Bostick, D. T. and Hercules, D. M., Anal. Chem. *47,* 447 (1975)
75. Auses, J. P., Cook, S. L. and Maloy, J. T., ibid. *47,* 244 (1975)
76. Williams III, D. C., Huff, G. F. and Seitz, W. R., ibid., *48,* 1003 (1975)
77. McCapra, F., Tutt, D. and Topping, R. M., in Lit. 1), p. 221
78. Bushnell, J., in Lit. 2), p. 319
79. Guthrie, J. P., Can. J. Chem. *52,* 2037 (1974)
80. Brolin, S. E., Wettermark, G. and Hammar, H., Strahlentherapie *153,* 124 (1977)
81. M.De Luca, M. A. (Ed.), Methods in Enzymology Vol. LVII, Academic Press New York 1978
82. Strehler, B. L. and Totter, J. R., Arch. Biochem. Biophys. *40,* 28 (1952)
83. Serio, M. and Pazzagli, M. (ds.), Luminescent Assays – Perspectives in Endocrinology and Clinical Chemistry, Serono Symposia Publications from Raven Press, Vol. 1, New York 1982
84. Wulff, K., Dt. Ges. f. Klin. Chem. *1981,* 135
85. McElroy, W. D., Hastings, J. W., Coulombre, J. and Sonnenfeld, V., Arch. Biochem. Biophys. *46,* 399 (1953)
86. Lundin, A. and Thore, A., Anal. Biochem. *66,* 47 (1975)
87. Lundin, A., Rickardson, A. and Thore, A., ibid. *75,* 611 (1976)
88. Wulff, K., Haar, H. P. and Michal, G., in Lit. 83), p. 47
89. Schram, E., in Lit. 83), p. 1
90. Kohen, F., Pazzagli, M., Kim, J. B., Lindner, H. R. and Boguslaski, R. C., FEBS Lett. *104,* 201 (1979)
91. Kohen, F., Kim, J. B., Lindner, H. R. and Eshar, Z., in Lit. 83), p. 169
92. Ewetz, L. and Thore, A., Anal. Biochem. *71,* 564 (1976)
93. Vasileff, T. P., Svarnas, G., Neufeld, H. A. and Spero, L., Experientia *30,* 20 (1974)
94. Riemann, B., in Lit. 1), p. 316.
95. Cairns, J. E., Nutt, S. G. and Afghan, K. B., in Lit. 1), p. 303
96. Lundin, A., Baltscheffsky, M. and Höijer, B., in Lit. 1), p. 339
97. De Luca, M., Wannlund, J. and McElroy, W. D. in 2) p. 179
98. Van de Werf, R. and Verstraete, W., in Lit. 1), p. 33
99. Sigalat, C. and Kouchovsky, Y., in Lit. 1), p. 367
100. Lundin, A. and Styrelius, I., Clin. Chim. Acta *87,* 199 (1978)
101. Witteveen, S. A. G., et al., Proc. Natl. Acad. Sci. *71,* 1384 (1979)
102. Tarkkanen, P., et al., Clin. Chem. *25,* 1644 (1979)
103. Wulff, K., et al., Fresenius Z. anal. Chem. *301,* 173 (1980)
104. Wulff, K., Stähler, F. and Gruber W., in Lit. 2), p. 209
105. Idahl, L. A., in Lit. 1), p. 401
106. Lundin, A., in Lit. 2), p. 187
107. Ulitzur, S. and Hastings, J. W., Proc. Natl. Acad. Sci. *75,* 266 (1978)
108. Ulitzur, S., in Lit. 1), p. 135
109. Ames, B. N., McCann, J. and Yamasaki, F., Mutation Res. *31,* 347 (1975)
110. Ulitzur, S., Weiser, I. and Yannai, S., in Lit. 2), p. 139
111. Agren, A., Berne, C. and Brolin, S. E., Anal. Biochem. *78,* 229 (1977)
111 a. Cilento, G., in: Adam, W. and Cilento, G., (Eds.) Chemical and Biological Generation of Excited States, p. 277, Academic Press, New York etc. 1982. Duran, N., ibid., p. 345
112. Allen, R. C., Stjernholm, R. L. and Steele, R. H., Biochem., Biophys. Res. Commun. *47,* 679 (1972)
113. Allen, R. C., Yevich, S. J., Orth, R. W. and Steele, R. H., Biochem. Biophys. Res. Commun. *60,* 909 (1974)

113a. Allen, R. C. and Loose, L. D., ibid. *69*, 245 (1976)

114. De Chatelet, L. R., Shirley, P. S. and Johnson jun., R. B., Blood *47*, 545 (1976)

115. Allen, R. C. in Lit. 2), p. 63

116. Allen, R. C., Photochem. Photobiol. *30*, 157 (1979)

117. Allen, R. C., in Lysosomes in Applied Biology and Therapeutics, Vol. 6, p. 197 (Eds.: Dingle, J. T., Jaques, P. J. and Shaw, I. H.), North Holland Publ. Co, Amsterdam 1979

118. Foote, C. S., in Lit. 2), p. 81

119. Abraham, G. F. (Ed.), Handbook of Radioimmunoassay, Marcel Dekker, New York, 1977 – for example

120. Woodhead, J. S., Weeks, I., Campbell, A. K., Ryall, M. F. T., Hart, R., Richardson, A. and McCapra, F., in Lit. 83), p. 147

121. Schroeder, H. R., Vogelhut, P. O. Carrico, R. J., Boguslaski, R. C. and Buckler, R. K., Anal. Chem. *48*, 1933 (1976)

122. Schroeder, H. R., Yeager, F. M., Boguslaski, R. C. and Vogelhut, P. O., J. Immun. Meth. *25*, 275 (1979)

123. Schroeder, H. R., in Lit. 83), p. 129

124. Schroeder, H. R. and Yeager, F., Anal. Chem. *50*, 1114 (1978)

125. Kohen, F., Kim, J. B., Barnard, G. and Lindner, H. R., Steroids *36*, 405 (1980)

125a. Barnard, G., Collins, W. P., Kohen, F. and Lindner H. R., lit. 2) p. 3

126. Pazzagli, M., Boelli, G. F., Messeri, G., Martinazzo, G., Tommasi, A., Salerno, R. and Serio, M., in Lit. 83), p. 191

127. Patel, A., Woodhead, J. S., Campbell, A. K., Hart, R. C. and McCapra F., in Lit. 83), p. 181.

128. Simpson, J. S. A., Campbell, A. K., Richardson, A., Hart, R. and McCapra, F., in Lit. 2), p. 673

128a. Weeks, I., Beheshti, I., McCapra, F., Campbell, A. K. and Woodhead, J. S., Clin. Chem. *29*, 1474 (1983)

129. Gundermann, K.-D., Wulff, K., Linke, R. and Stähler, F., in Lit. 83), p. 157

129a. Patel, A., Davies, C. J., Campbell, A. K. and McCapra, F., Anal Biochem. *129*, 162 (1983)

130. Wannlund, J., Azari, J., Levine, L. and de Luca, M., Biochem. Biophys. Res. Commun. *96*, 440 (1980)

131. DeLuca, M., in Lit. 83), p. 115

132. U.S. 3.576.987; U.S. 3.597.362 (American Cyanamid) (1971)

133. U.S. 3.969.263 (1975); American Cyanamid

134. U.S. 3.704.231 (1966); American Cyanamid

135. U.S. 3.769.227 (1968)

136. U.S. 3.824.289 (1971)

137. U.S. 3.888.785 (1970)

138. U.S. 4.086.183 (1976)

139. U.S. 3.496.111 (1966)

140. U.S. 3.679.594 (1966)

141. U.S. 3.697.434 (1965)

142. U.S. 3.502.588 (1966)

143. U.S. 3.616.593 (1969); U.S. Navy

144. U.S. 3.714.054 (1965)

145. U.S. 4.089.797 (1977)

146. U.S. 3.629.129 (1967)

147. U.S. 3.576.753 (1967)

148. U.S. 3.576.752 (1967)

149. U.S. 3.551.343 (1968)
150. U.S. 3.551.342 (1967)
151. DT 2.249.867 (1971)
152. SU 531.840 (1974)
153. SU 691.478 (1977)
154. J 54.046.184 (1977)
155. FR 2.043.849 (1969)
156. DT 2.533.779 (1974)
157. U.S. 815.090 (1977)
158. Babko, A. K., in: Preprints of the International Symposium on Chemiluminescence, Durham, N. C., p. 429 (1965)
159. Journal of Bioluminescence and Chemiluminescence, Ed.-in-Chief L. J. Kricka, Birmingham, U. K., John Wiley & Sons, London, New York
160. Hastings, J. W., Baldwin, T. O. and Nicoli, M. Z., in: Methods in Enzymology 57 (M. de Luca ed.), p. 135 (1978)
161. See also Slawinska, D. and Slawinski, J., in: Chemi- and Bioluminescence (J. G. Burr, ed.), p. 495, M. Dekker, Inc., New York, Basel 1985

XIV. Instrumentation

The growth of interest in the analytical applications of chemiluminescence has resulted in a corresponding amount of attention being paid to the need for convenient and sensitive instrumentation. Monographs [1, 2] dealing with such applications are useful sources of information.

Although the descriptions of light emission still abound with subjective statements such as "bright", "weak" and "easily visible", there are very many ways in which a quantitative result can be obtained. Absolute intensity measurements are, however, seldom performed. Most quantum yields are reported relative to one or other of the available standards [3, 4]. Chemiluminescent spectra are routinely produced by fluorescence spectrometers with the excitation beam switched off or obscured. For very weak spectra, filter systems rather than true dispersive (e. g. grating) spectrometers are often employed [3]. It has been estimated in a recent review [5] that some 14% of the full papers in "Analytical Chemistry" include the techniques of phosphorescence, fluorescence or chemiluminescence.

The earliest use of a photoelectric device was reported by Weber and co-workers in 1942 [6], employing a selenium cell. With modern electronic components it is very easy to build simple and cheap circuits for use with inexpensive photomultipliers [7]. A block diagram [7] of the most basic apparatus is shown in Fig. 29.

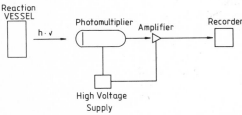

More recently their has been considerable development of photon counting and digital methods (Fig. 30 [7])

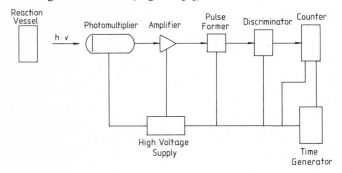

which are necessary for the measurement of the weakest emissions. A welcome addition to the tools available is provided by easily accessible micro-computers and micro-processors. These have been used to control the addition of reagents [8, 9], and in the control of the measurements in general [10, 11], Flow systems can be of assistance in the detection of weak signals, such as that from diazaquinone chemiluminescence [12, 13]. The problems of the measurement of weak emission are discussed in many of the references in this section.

It is unnecessary to provide further detail of the apparatus since some excellent descriptions are in the literature [14–20]. These allow the construction of simple and hence flexible systems, yet thoroughly examine the theory of light measurement so that the investigator can choose the most suitable apparatus. With the growth of interest in luminescence-based immunoassay methods, there has been a rush by many manufacturers to produce luminometers to serve this potentially lucrative market. The instruments and the addresses of the companies are listed in reference 2. It is important to point out however, that these instruments are rarely suitable for research work in the field of chemiluminescence itself. They are sensitive and where the analytical requirements are well defined, they can be very convenient. More flexibility is required in the apparatus used for investigations of the fundamental light reaction and its mechanism.

XIV.1 References

1. Methods in Enzymology, Vol. LVII (1978), Ed. DeLuca, M., Academic Press, New York
2. Kricka, L. J. and Carter, T. J. N., Clinical and Biochemical Luminescence. (Vol. 12 of Clinical and Biochemical Analysis, Marcel Dekker Inc New York and Basel 1982)
3. Lee, J. and Seliger, H. H., Photochem. Photobiol. *4,* 1015 (1965) Seliger, H. H., in ref. 1, p. 560
4. Oikari, T. E. T., Hemmila, I. A. and Soini, E. J., in Analytical Applications of Bioluminescence and Chemiluminescence (Kricka, L. J., Stanley, P. E., Thorpe, G. H. G. and Whitehead, T. P., eds.) Academic Press, New York, 1984, p. 475
5. Wehry, E. L., Anal. Chem. *54,* 131 R (1982)
6. Weber, K., Rezek, A. and Vouk, V., Ber. dt. chem. Ges. *75,* 1141 (1942)
7. Wulff, K., Mitt. Dt. Ges. f. Klin. Chemie *1981,* 135
8. Stieg, S. and Nieman, T. A., Anal. Chem. *52,* 796 (1980)
9. Balciunas, R., Holler, F. J., Notz, P. K., Johnson, E. R., Rothman, I. D. and Crouch, S. R., ibid. *53,* 1484 (1981)
10. Stieg, S. and Nieman, T. A., ibid. *53,* 800 (1981)
11. Marino, D. F. and Ingle, J. D., ibid. *53,* 1175 (1981)
12. Gundermann, K.-D., Unger, H. and Stauff, J., J. Chem. Res. (S) *1978,* 318, *M 1978,* 3846
13. Seitz, W. R., Proc. Int. Sympos. Anal. Appl. Biolumin. and Chemilum. *1979,* 683 (C. A. *93,* 40790 (1981))
14. Seliger, H. H., see Lit. 10)
15. Stanley, P. E., in Liquid Scintillation Counting (Crook, M. A. and Johnson, P. Eds.) p. 253, Heyden, London 1974
16. Krinsky, N. I., in Wassermann, H. H. and Murray R. W. (Eds.) Singlet Oxygen, p. 597, esp. p. 604

17. Berthold, F. and Kubisiak, H., in Serio, M. and Pazzagli, M. (Eds.) Luminescent Assays, Perspectives in Endocrinology and Clinical Chemistry, Serono Symposia Publications from Raven Press, Vol. 1., p. 23. Raven Press, New York, 1982
18. Wampler, J. E., in Bioluminescence in Action (Herring, P. J., ed.) Academic Press, London, 1978 p. 1
19. Van Dyke, K., Bioluminescence and Chemiluminescence – Instruments and Applications, CRC Press, Boca Raton, Florida, 1985
20. Wampler, J. E. in Chemi- and Bioluminescence, (Burr, J., ed.) Dekker, New York, 1985 p. 1

XV. Chemiluminescent Demonstrations [1]

Demonstrations of scientific phenomena have a long and honourable history and the more dramatic pyrotechnic examples of chemical reactions are deservedly popular. However the gentle glow from variously coloured chemiluminescing solutions has a unique appeal. Sufficient intensity is sometimes obtainable for the demonstrations to be visible in daylight, but their appearance in all the colours of the spectrum in a totally darkened room is a memorable sight.

By reference to the appropriate sections of this book they can be used to demonstrate normally difficult to illustrate concepts such as quantisation, radical ion recombination, electron transfer, energy transfer, temperature effects on reactions and thermodynamics. Interesting discussions on their relationship to photosynthesis, vision and photochemistry can be provoked.

Even although many of the compounds can be purchased (cf. the reference section for sources), most are readily synthesised and can make a very useful training exercise in student laboratories, with an enticing end result.

Luminol

The first published demonstration [2] used luminol, and E. H. White's later description [3] still serves as the basis for some of the best demonstrations. Using this compound a bright blue, short-lived emission of light is obtained in aqueous solution using transition metals, usually copper, iron or hemin, as catalysts. A still brighter, and much more sustained blue-green light results when a solution of luminol in DMSO or DMF is treated with strong base.

1 Two solutions are required. For the first, luminol (0.1 g) is dissolved in H_2O (500 ml) by the addition of Na_2CO_3 (2.0 g) and shaken. The solution is buffered by adding $NaHCO_3$ (12.0 g). Copper sulphate (0.3 g) is then added as the catalyst. This solution is stable enough for use in demonstrations but should not be stored for more than a day or two.

   The second solution contains H_2O_2 (5 ml of 30% solution per litre of H_2O). Mixing in roughly equal amounts gives bright blue light lasting for about 2 minutes. Pouring both solutions simultaneously into a large beaker, allowing the streams to mix from a height, gives good results.

   This reaction can be used to demonstrate energy transfer by adding fluorescen or rhodamine B (about 20 mg of each on the scales given in 10 ml. H_2O and ethanol respectively). However, lucigenin (see later) is much more suitable for this purpose since it lasts longer.

2 Luminol (0.05 to 0.1 g) is added to a 1-litre flask with a very good rubber or ground glass stopper containing a 1 cm layer of KOH pellets just covered by DMSO. Vigorous shaking leads to a gradual build up of light. At the higher concentrations this may take some time but it is possible with practice to guess a sudden increase in light intensity, and supply a suitable commentary during the

demonstration. If left undisturbed, this reaction is still bright on shaking over 12 hours later. The bright meniscus demonstrates the reaction with O_2 by diffusion. Light emission occurs more readily, but less interestingly by using potassium *tert*-butoxide (1 g) in DMSO (100 ml) and luminol (0.05 g).

Lucigenin
Although many of the procedures for demonstrating this readily available compound use NaOH as the base, a much cleaner and more convenient reaction is obtained as follows.

A saturated solution of lucigenin in ethanol is prepared by shaking it at room temperature. It will keep for many months. For the demonstration H_2O_2 (1 ml of a 30%–100% vol-solution) is added to 100 ml of lucigenin solution just before use (it is however useable over several hours). On the addition of an equal volume of ethanol containing 20% v/v of 0.88 NH_4OH a blue green emission is obtained lasting for about 15 minutes.

The ease with which this solution is prepared, and its simplicity suit it to the demonstration of "chemical sign writing". An effective demonstration is obtained by preparing the words "cold light" on a white board in 1 cm glass tubing, fed by a funnel and rubber tubing about 0.5 m above. Two dropping funnels supply the funnel, and a tap and flask at the back of the board collect the effluent. Addition to the funnel of fluorescein (as a solution of 1 g in 50 ml dilute NaOH) and rhodamine B (1 g in 50 ml EtOH) in about 5 ml amounts changes the colour as the solution flows through the tube. Closing all taps gives a bright display for about fifteen minutes.

Active Oxalate Esters
This patented system [4] is the only one to achieve commercial acceptance and is widely used throughout the world for emergency and recreational lighting. The Cyalume® light-sticks are well developed formulations, containing hydrogen peroxide and catalyst in *tert.*-butyl alcohol in an alkyl phthalate in the outer tube. One litre produces the equivalent of a 40 W bulb burning for over ninety minutes. The separate materials can be taken from the tube for special demonstrations, but they are expensive. Other published methods, while not giving quite the efficiency of the very well designed Cyanamide system, give the chemist an opportunity to impress in quantity!

The commercial system uses the oxalate (1)

XV (1)

The more readily available oxalates (2) and (3) give adequate light

XV (2)   XV (3)

emission. The bis-dinitrophenyl oxalate (3) does not require a catalyst and provides the simplest demonstration. The bis-trichlorophenyl oxalate is slower reacting and gives a lesser intensity, but a much longer lifetime. Suitable uncatalysed mixtures can be prepared to suit a variety of occasions. Alternatively the reaction of the trichloro-derivative can be catalysed by sodium salicylate as described below. The synthesis of both of these compounds has been described [5]. It is assumed that the high efficiency obtained by the commercial preparation by using almost anhydrous (90%) H_2O_2 is easily foregone in the interests of safety, a 30% solution of H_2O_2 being quite effective.

Sensitisers
Many of the classical fluorescers such as fluorescein are unsatisfactory in the oxalate reaction and a range of aromatic hydrocarbons is preferred. Diphenylanthracene (blue) and 9,10-Bis-phenethynyl anthracene (green) are easily made and others such as rubrene (yellow-orange) and violanthrone (orange-red) are commercially available.

Demonstration
(a) *Using bis-(2,4-dinitrophenyl)-oxalate*
The amounts given are convenient, rather than optimum. Higher quantum yields can be obtained by using published methods, but the difference is not significant in a lecture demonstration. Nearly anhydrous (90%) H_2O_2 is equally unnecessary, although again the quantum yield could be increased at the expense of safety.
Hydrogen peroxide (1 ml, 30% aqueous solution) is dissolved in *tert.*-butanol (10 ml) and added to a dialkyl phthalate such as diethylphthalate (20 ml). A fluorescer such as diphenyl-anthracene (0.1 g) is dissolved in a second quantity of dialkyl phthalate (100 ml) together with bis-(2,4-dinitrophenyl)-oxalate (from 0.1–10 g). The oxalate is not very soluble, but the suspension dissolves as it reacts. The commercial oxalate (1) is designed to achieve concentrations as high as 600 g/litre. The dinitrophenyl oxalate also quenches at high concentrations and nothing is to be gained in this case at the higher end of the concentration range. Use of a catalyst is unnecessary and is in fact deleterious.
(b) *Using bis-(2,4,6-trichlorophenyl)-oxalate*
This can be used as above. It is more soluble and less liable to cause concentration quenching. However a basic catalyst is needed to produce a more rapid build-up of light intensity. If the catalyst is used, mixing with the previous solution is inadvisable. The uncatalysed solution, mixed with (a), can be arranged to extend the lifetime of the glow, while retaining the immediate impact of procedure (a). If the trichlorophenyl compound is used on its own, dissolve sodium salicylate (0.02 g) in the *tert.*-butanol-hydrogen peroxide solution before mixing.
White light can be produced by mixing glowing solutions of blue (diphenyl-anthracene) and yellow (rubrene) formulations.

TMAE (Tetrakis-dimethylaminoethylene)

Although this compound is easily synthesised [5] it is not very expensive and can be purchased. It is of course the easiest of all chemiluminescent materials to demonstrate – merely taking the top off the bottle causes it to glow! It can be demonstrated by transferring some material to a stoppered flask and spreading it along the sides by turning the flask slowly. However this is not as impressive as the reactions previously described if treated in this way. Its main advantage is that it can be used in a "magic light pen". This can be constructed in various ways. The simplest is obtained by using a 1 cm diameter test-tube. A tightly rolled 2 cm wide strip of felt is wedged in the neck with sufficient felt protruding to produce a felt pen. The tube is charged through the felt using a glass hypodermic syringe. Use only enough for the immediate purpose since O_2 must be excluded if TMAE is to be kept.

The whole test-tube could of course be stored in a vial only very slightly larger than the test-tube. The small amount of O_2 is soon consumed and enough TMAE can be left for several occasions. More elegant variations can be constructed in which a properly stoppered glass container is itself the pen, allowing it to be used as long as the supply of TMAE lasts.

Acridinium esters

These react extremely quickly with alkaline hydrogen peroxide in ethanol. The resulting flash of light can be startlingly bright.

Synthesis

Acridine 9-carboxylic acid can be purchased (Aldrich) but large quantities are economically prepared as described below.

Acridine-9-carboxylic acid

Acridine (25 g, 0.14 M) is dissolved in ethanol (110 ml) and the solution added to glacial acetic acid (9 ml). A solution of KCN (12 g, 0.31 M) in 20 ml water is added and the mixture stirred under reflux for 1 hour. After cooling, the precipitate is filtered off and washed with 2 M NaOH solution and water. It is dissolved in chloroform and dried over anydrous Mg_2SO_4. Evaporation of the filtrate to dryness and recrystallisation from n-propanol yields acridine 9-nitrile (19.6 g 95 mM, 70%), (m.p 185–186°, I. R. 2222 cm^{-1}, $-C\equiv N$).

The acridine-9-nitrile (15 g, 70 mM) is added to conc. sulphuric acid (120 ml) and heated on a steam bath for 2.5 hours. It is cooled to 0 °C in an iced water bath and sodium nitrite (55 g, 7.5 M) is slowly added so that brown fumes are evolved. The mixture is then *carefully* heated with a bunsen burner until no more gas is given off. It is then stirred at 100 °C for 2 hours, cooled and slowly poured into iced water to precipitate the yellow product. This is filtered off, washed with water and sucked dry. The product is then dissolved in the minimum of 2 M NaOH solution and filtered through a sintered funnel. The deep red solution is treated with conc. HCl until the yellow precipitate has permanently reformed. This is filtered, washed with water and sucked dry. It is further dried at 55 °C under reduced pressure for 24 hours (14.7 g, 65 mM, 89%), (i.r 1650 cm^{-1}, acid C=O).

Acridine-9-carbonyl chloride

Acridine-9-carboxylic acid is refluxed in a large excess of redistilled thionyl chloride under dry conditions until solution is complete (ca. 3 hours). The reagent is evaporated off (any residue of thionyl chloride being removed by co-evaporation with dry benzene) to give a yellow powder that is used without further purification (IR $1780\,cm^{-1}$ $Cl-C=O$).

Phenyl acridine 9-carboxylate

Acridine-9-carbonyl chloride is stirred in dry pyridine forming a brown suspension after a few minutes. Phenol (in about 10% molar excess) is dissolved in a few ml of pyridine and added to the suspension. The reaction mixture is stirred for 8–12 hours at room temperature and is then poured into a stirred mixture of ice and conc. hydrochloric acid to precipitate the product. This is filtered, washed with water and sucked dry. Recrystallisation gives the pure product in 85% yield.

Recrystallisation from toluene gave beige needles, m. p. 186–188°. NMR δ 7.30–8.42 – complex aromatic pattern; IR $1740\,cm^{-1}$.

Phenyl 10-methyacridinium-9-carboxylate methosulphate

Phenylacridine-9-carboxylate (500 mg, 1.7 mM) is dissolved in 40 ml dry toluene. Dimethyl sulphate (4 ml, 42 mM) is added and after stirring at 100° for 5 hours the cooled precipitated salt is filtered and washed with toluene to dissolve excess dimethyl sulphate. The solid is then washed with diethyl ether and allowed to suck dry. The product is a bright yellow powder (635 mg, 89%). Recrystallisation from ethanol affords a pure sample.

A reproducible melting point could not be obtained, a situation similarly encountered by previous observers, and so it would appear that salts of this nature do not have a reproduceable melting point.

NMR (d_6-DMSO)

	δ 4.97 (3 H,s)	$N-CH_3$
	δ 7.42–9.07 (12 H,m) aromatic H's	
IR	$1608\,cm^{-1}$ aromatic $C=C$	
	$1760\,cm^{-1}$ ester $C=O$	
MS	315 (M^++H), 299 (M^++H-CH_3), 206 ($299-$)C_6H_5,	
	178 ($206-CO$)	

Phenyl 10-methylacridan-9-carboxylate

The acridinium salt (290 mg, 0.70 mM) is dissolved in 20 mls of distilled methanol. The clear, yellow solution is acidified with two drops of HCl (2 M), the catalyst (25 mg) added and hydrogen bubbled into the stirred mixture at room temperature. After about 10 minutes the solution turns an intense purple colour which completely disappears to leave a suspension of the white product and the black catalyst. Chloroform (20 ml) is added to dissolve the acridan and the solution is filtered, quickly washed with water (5 ml) and evaporated at room temperature under vacuum to dryness to give an off-white solid (16 mg, 72%). Recrystallisation from ethanol gave white needles.

Mp   116–118 °C

δ   3.42 (3 H,s)   N–CH$_9$
δ   5.17 (1 H,s)   C$_3$–H
δ   6.84–7.40 (13 H,m) aromatic H's

Alternative synthesis of 10-methylacridan-9-carboxylate
Phenyl acridinium 9-carboxylate (1 g) ist refluxed with zinc dust (0.3 g) in acetic acid solution (10 ml) until the intermediate purple colour vanishes. Any unreacted zinc dust is removed by filtration, and the product precipitated by addition of water. If the precipitate is difficult to filter, extraction with ether followed by drying over MgSo$_4$ and evaporation gives good recovery. Recrystallisation from ethanol gives white needles m. p. 116–118 in 60% yield.

Demonstration of Acridinium and Acridan esters.
Although these compounds react with quantum yields greater than luminol and less than that of the active oxalate esters, they are unsurpassed in producing very high intensities, albeit for a short time.

(a) A small quantity of phenyl acridinium methosulphate as a saturated solution (or indeed suspension) in acetone is added dropwise to 100 ml of 2 N ethanolic NaOH solution to which has been added 5 ml of 30% (100 vol.) hydrogen peroxide. Bright flashes result.

(b) A solution of KOH (1 g) in distilled water (100 ml) is freshly prepared and added to a 1 litre measuring cylinder and the cylinder filled almost to the top with industrial grade ethanol or methylated spirit. Hydrogen peroxide (10 ml, 30% solution) is added shortly before the demonstration and the solution is mixed. Phenyl acridinium ester is then added as a coarse powder (or small crystals) to give an entrancing luminescent "snowstorm" effect. After addition of about 50–100 mg of material, the cylinder is stoppered or covered by hand, and rapidly inverted. In a totally darkened room the flash is impressive.

(c) The phenyl acridan can be added in very small amounts to either dimethyl sulphoxide or dimethyl formamide with a single layer of KOH pellets covering the bottom of a small conical flask. Rapid shaking produces probably the brightest display in chemiluminescence. Since these are excellent mechanistic models for the coelenterate and firefly luciferins, the value of the demonstration is enhanced. It is possible to see the luminescent meniscus which forms on standing, when there is still unconsumed acridan, thus demonstrating the ready reaction with oxygen, one of the steps in the biological reaction (see the discussion of bioluminescence, Chap. XIII).

Chemiluminescent Clock Reactions
Clock reactions, in which a kinetic sequence is so balanced as to lead to long induction periods followed by extremely rapid catalytic activity are well known in chemical systems. Luminol is used in two luminescent clock reactions [3, 7]. In the first, the oxidation of cyanide ion by H$_2$O$_2$ is required before the Cu-(II) can catalyse the luminol reaction. The second is the safer, in that harmless cysteine is the inhibitor which is oxidised.

Three solutions are required. A, B and C and variations in the amount of C give changes in induction time.

Solution A
DL-Cysteine hydrochloride (0.08 g) is dissolved in H_2O (50 ml). This should be freshly made up as cysteine oxidises in solution in air.

Solution B
Luminol (0.1 g) is dissolved in H_2O (100 ml) containing 5.0 g of NaOH.

Solution C
$CuSO_4 \cdot 5H_2O$ (0.06 g) in H_2O (1 L)

Demonstration
In each of three 100 ml conical flasks (set 1) place solution A and solution B (5.5 ml).

In another three flasks (set 2) separately place 30 ml, 40 ml and 50 ml of solution C.

Dilute 10 ml of H_2O_2 (30% solution) to 100 ml and place 0.5 ml of this solution in each of the flasks of set 2.

As simultaneously as possible mix the contents of the flasks of set 1 and set 2. After an induction period of about 30 seconds, a brief glow will appear in the flask containing 50 ml of solution C followed by the others, with successive 10 second intervals.

The induction period is proportional to the cysteine concentration and inversely proportional to the $CuSO_4$ concentration, so induction can be varied, although predictability becomes difficult if the range suggested were greatly altered. Practice is strongly recommended!

The well known high affinity of thiols for Cu-II inhibits its catalytic effect by forming a complex, sufficient Cu-II being released only on oxidation of the thiol group.

XV.1 References and Sources of Materials

1. Shakhashiri, B. Z., Chemical Demonstrations, Vol. 1, University of Wisconsin Press 1983
2. Huntress, E. H., Stanley, L. N. and Parker, A. S., J. Chem. Educ. *11,* 142 (1934)
3. White, E. H., ibid. *34,* 275 (1957)
4. Rauhut, M. M., Kirk-Othmer-Encyclopedia of Chemical Technology Vol. 5, (3rd edition) pp. 416–450 (1979)
5. Rauhut, M. M., Acc. Chem. Res. *2,* 80 (1969); Rauhut, M. M., Bollyky, J. J., Roberts, G. B., Loy, M., Whitman, H. and Iannotta, A. V., J. Amer. Chem. Soc. *89,* 6515 (1967)
6. Winberg, H. E., Downing, J. R. and Coffman, D. D., ibid. *87,* 2054 (1965)
7. Young, K. E., J. Chem. Educ. *51,* A 122 (1974)

Sources

It is assumed that common reaction intermediates are readily available to would-be demonstrators. Luminol, acridine, TMAE, diphenylanthracene, rubrene, fluorescein, rhodamine B are all obtainable from the Aldrich Chemical Company, which has branches in most countries. Violanthrone can be obtained from ICN Pharmaceuticals Inc. K and K Labs Division, 121, Express Street, Plainview, New York 11803, USA.

XVI. Chemiluminescence in the Future

The last twenty years have seen most of the mystery attached to the beguiling emission of light from organic compounds in solution resolved. Nevertheless although there are several established mechanisms to assist the investigator of new chemiluminescent reactions, very few indeed can be said to be fully understood. This is perhaps not too different from the case in organic reaction mechanism generally, but chemiluminescence research includes the extra dimension of the excited state. Increased uncertainty is inevitable when the gap between the language used by chemists to describe organic reactions and that of quantum mechanics has to be bridged. The progress made in the recent past is satisfying, but there is a little unease about even the best understood examples, and settling these points is one task for the future. The CIEEL mechanism much discussed in this book covers much ground, but further work is needed particularly in the intramolecular cases before we can be convinced that discrete electron transfer steps are necessarily present. The many examples in which small changes of structure result in puzzling differences in the multiplicity (singlet vs triplet) of the excited products also requires more investigation.

There is also a long list of reactions which do not have a convincing chemical mechanism, so that any proposed excitation mechanism sits uneasily on them. Foremost among these is luminol, and although readers will no doubt be able to add other examples, tetrakis-dimethylaminoethylene, models for bacterial luminescence and the whole range of very weakly luminescent reactions involved in phagocytosis and polymer degradation should be included. In the background, at least from the standpoint of the present volume, are several highly intriguing bioluminescent reactions in organisms such as bacteria, fungi, Latia, dinoflagellates, euphausid shrimp and certain squid. There is sufficient information as to suggest that new categories of chemiluminescent reactions may have to be devised as part of the explanations for these light reactions.

The pure chemical aspect must not be forgotten: without an established knowledge of the molecules involved in chemi- and bioluminescence and without the capacity to synthesize them there will be no real progress in this fascinating field.

Lastly, there are the applications of chemiluminescence. The oxalate esters give us a satisfactory source of cold light for illumination. Yet, as the inventors of this system have calculated, only a few percent of the light theoretically available in the ideal chemiluminescent reaction has been obtained. The analytical uses are potentially the most valuable. Here the situation is poised between classical analysis such as that for trace metals and the much more subtle use in biological systems, particularly in immunoassay. Excellent assays now exist, but chemiluminescence has many competitors in the field. Its success is highly dependent on market requirements, some of which have still to be determined. Many new research workers are now in the field, particularly in applications, and we can expect consolidation of the gains made here together with a few delightful surprises in our as yet incomplete mechanistic picture.

Appendix

Table of Luminol Type Hydrazides

Table 14. Luminol Type Hydrazides

Phthalic hydrazide 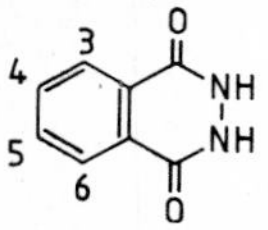	Chemilu-minescence $[\lambda_{max}]$ (nm)	$\varnothing_{CL}$ (luminol = 100*) H_2O^a	DMSO^b	Ref.
3-(β-Dialkylamino vinyl)-				
3- (i–C_3H_7)$_2$ N–CH = CH –	495^c	<0,8	7,0^c	VII.7. [47]
3- (i–C_4H_9)$_2$ N–CH = CH –	495^c	<0,8	35,0^c	VII.7. [47]
3- (i–C_5H_{11})$_2$ N–CH = CH –	495^c	<0,8	20,0^c	VII.7. [47]
3- (p-subst.) Phenyl-				
3-p- 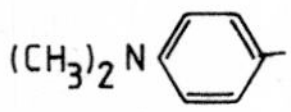 (CH$_3$)$_2$ N	530–540^d	0,75	2,0	VII.7. [28]
3-p- 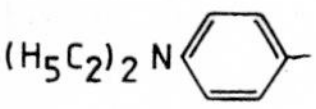 (H$_5$C$_2$)$_2$ N	460–490^d	0,30	4,0	VII.7. [28]
3,6-disubstituted phthalhydrazides				
3-amino-6-methyl	405	187	–	VII.7. [9]
3-amino-6-ethyl				VII.7. [30]
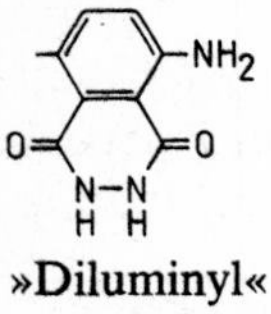 3-amino-6-	510	6		VII.7. [28]
	450^d		30	

»Diluminyl«

* Corresponding to the luminol standard of Lee and Seliger, VII.7. [22] = 1.25×10^{-2}.

Hydrazide	Chemilu- minescence [λ_{max}] (nm)	$\emptyset_{CL}$		Ref.
		H_2O^a	$DMSO^b$	
4-substituted *phthalhydrazides*				
4-$(CH_3)_2N$ –	437	50[e]	–	VII.7. [44, 9]
4-$(C_2H_5)_2N$ –	439	63[e]	–	VII.7. [44, 9]
4-$(n–C_4H_9)_2N$ –	442	93	–	VII.7. [9]
4-$(n–C_7H_{15})_2N$ –	439	25	–	VII.7. [9]
Naphthalene-1.2- **dicarboxylic acid-** **hydrazides**				

(structure of naphthalene-1,2-dicarboxylic acid hydrazide with positions 3,4,5,6,7,8)

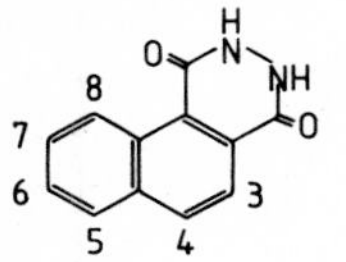 Hydrazide	Chemilu- minescence [λ_{max}] (nm)	$\emptyset_{CL}$ H_2O^a	$DMSO^b$	Ref.
5 – OH	550	very weak	164	VII.7. [97]
6 – OH	485	86	182	VII.7. [97]
7 – OH	512	120	301	VII.7. [97]
5 – $N(CH_3)_2$	550 485	7	2	VII.7. [97]
6 – $N(CH_3)_2$	485 450	150	14	VII.7. [97]
7 – $N(CH_3)_2$	514	230	60	VII.7. [15]
Anthracene – *2.3-dicarboxylic* *hydrazide*				

(structure of anthracene-2,3-dicarboxylic hydrazide with positions 1,2,3,4,5,6,7,8,9,10)

Hydrazide	Chemilu- minescence [λ_{max}] (nm)	$\emptyset_{CL}$ H_2O^a	$DMSO^b$	Ref.
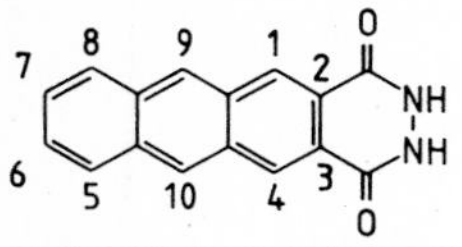	430	25	3,3	VII.7. [98]
6 – OCH_3	457	75	47	VII.7. [98]
6 – $N(CH_3)_2$	470	50	–	VII.7. [98]

Hydrazide	Chemiluminescence [λ_{max}] (nm)	$\varnothing_{CL}$		Ref.
		H_2O^a	$DMSO^b$	
9 – Br	–	–	5,3	VII.7.
9,10 – Diphenyl	450		30	VII.7.

Phenathrene –
– dicarboxylic
hydrazides

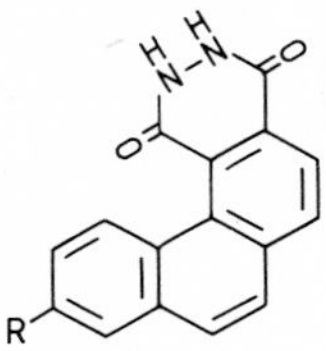				
R = H	368	–	–	VII.7. [28]
R = NH$_2$	490	2,1	–	VII.7. [28]
R = OCH$_3$	480	26	–	VII.7. [28]
R = OH	525	44	–	VII.7. [28]
R = (CH$_3$)$_2$N	520	115	–	VII.7. [28]
R = (C$_2$H$_5$)N	510	129	–	VII.7. [28]

Fluorene-1,2-dicarbox-
ylic acid hydrazides

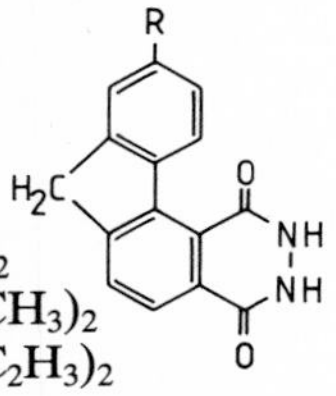				
R = NH$_2$	450	0,03	–	VII.7. [28]
R = N(CH$_3$)$_2$	450	0,10	–	VII.7. [28]
R = N(C$_2$H$_3$)$_2$	500	0,15	–	VII.7. [28]

Benzo[g,h,i]perylene-1.2-
dicarboxylic hydrazide

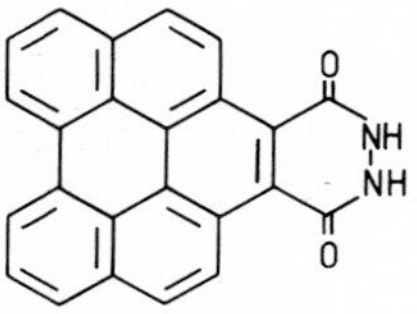	429,447	–	580[f]	VII.7. [20]

| Hydrazide | Chemilu-minescence $[\lambda_{max}]$ (nm) | $\varnothing_{CL}$ | | Ref. |
		H_2O^a	DMSO^b	
Coronene-1,2-dicarboxylic acid hydrazide				
	437,461	–	360	VII.9. [16]

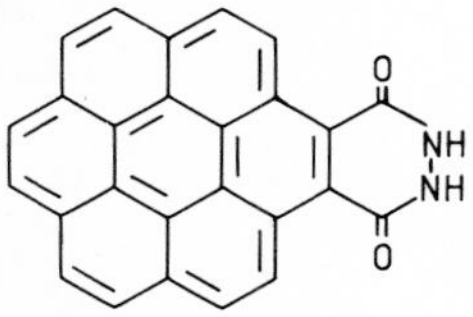

a) Alkali / H_2O_2/hemin
b) DMSO/K *tert*-butoxide/O_2
c) as 40–45% DMSO/60–55% H_2O
d) in DMSO
e) Even small changes in reaction conditions (especially pH) give rise to changes in $\varnothing_{CL}$; e.g. for 4-dimethylamino- and 4-diethylaminophthalhydrazide, the relative chemi-luminescence light yield is 87 and 120, respectively (luminol = 100), at pH 12–13 (VII.9. [55]).
   N-Dialkylamino- derivatives of phthalhydrazide and of naphthalene-dicarboxylic acid hydrazides were found to have their highest $\varnothing_{CL}$ at an alkalinity of 0.01 n NaOH (Ref. VII.9. [12]).
f) 30 mole % water/70 mole % DMSO, base, and oxygen.

Subject Index

Abbreviations: CL = chemiluminescence, BioL = bioluminescence

Notice: Only a few Luminol type hydrazides are presented separately in this subject index. For numerous other such hydrazides see Appendix, p. 205